VOLUME	EDITOR-IN-CHIEF	PAGES	
25	The Late Werner E. Bachmann	120	*Out of print*
26	The Late Homer Adkins	124	*Out of print*
27	R. L. Shriner	121	*Out of print*
28	H. R. Snyder	121	*Out of print*
29	Cliff S. Hamilton	119	*Out of print*

Collective Vol. 3 A revised edition of Annual
        Volumes 20–29
        E. C. Horning, *Editor-in-Chief*   890

30	The Late Arthur C. Cope	115	*Out of print*
31	R. S. Schrieber	122	
32	Richard T. Arnold	119	*Out of print*
33	Charles C. Price	115	*Out of print*
34	William S. Johnson	121	*Out of print*
35	T. L. Cairns	122	*Out of print*
36	N. L. Leonard	120	
37	James Cason	109	*Out of print*
38	John C. Sheehan	120	
39	Max Tishler	114	

Collective Vol. 4 A revised edition of Annual
        Volumes 30–39
        Norman Rabjohn, *Editor-in-Chief*   1036

40	Melvin S. Newman	114	
41	John D. Roberts	118	*Out of print*
42	Virgil Boekelheide	118	*Out of print*
43	B. C. McKusick	124	*Out of print*
44	William E. Parham	131	
45	William G. Dauben	118	
46	E. J. Corey	146	
47	William D. Emmons	140	
48	Peter Yates	164	
49	Kenneth B. Wiberg	124	*Out of print*

Collective Vol. 5 A revised edition of Annual
        Volumes 40–49
        Henry E. Baumgarten,
          Editor-in-Chief   1234

50	Ronald Breslow	110
51	Richard E. Benson	147
52	Herbert O. House	139
53	Arnold Brossi	129
54	Robert E. Ireland	

NO LONGER PROPERTY OF
FALVEY MEMORIAL LIBRARY

ORGANIC SYNTHESES

ORGANIC SYNTHESES

AN ANNUAL PUBLICATION OF SATISFACTORY
METHODS FOR THE PREPARATION
OF ORGANIC CHEMICALS

VOLUME 54

1974

ADVISORY BOARD

C. F. H. ALLEN
RICHARD T. ARNOLD
HENRY E. BAUMGARTEN
A. H. BLATT
VIRGIL BOEKELHEIDE
RONALD BRESLOW
T. L. CAIRNS
JAMES CASON
J. B. CONANT
E. J. COREY
WILLIAM G. DAUBEN
WILLIAM D. EMMONS
ALBERT ESCHENMOSER
L. F. FIESER
R. C. FUSON
HENRY GILMAN
C. S. HAMILTON
W. W. HARTMAN
E. C. HORNING

JOHN R. JOHNSON
WILLIAM S. JOHNSON
N. J. LEONARD
B. C. McKUSICK
C. S. MARVEL
MELVIN S. NEWMAN
C. R. NOLLER
W. E. PARHAM
CHARLES C. PRICE
NORMAN RABJOHN
JOHN D. ROBERTS
R. S. SCHREIBER
JOHN C. SHEEHAN
RALPH L. SHRINER
H. R. SNYDER
MAX TISHLER
KENNETH B. WIBERG
PETER YATES

BOARD OF EDITORS

ROBERT E. IRELAND, *Editor-in-Chief*

RICHARD E. BENSON
ARNOLD BROSSI
GEORGE H. BÜCHI
HERBERT O. HOUSE

CARL R. JOHNSON
SATORU MASAMUNE
WATARU NAGATA
ZDENEK VALENTA

WAYLAND E. NOLAND, *Secretary to the Board*
University of Minnesota, Minneapolis, Minnesota

FORMER MEMBERS OF THE BOARD, NOW DECEASED

ROGER ADAMS
HOMER ADKINS
WERNER E. BACHMANN
WALLACE H. CAROTHERS
H. T. CLARKE

ARTHUR C. COPE
NATHAN L. DRAKE
OLIVER KAMM
LEE IRVIN SMITH
FRANK C. WHITMORE

JOHN WILEY AND SONS
NEW YORK · LONDON · SYDNEY · TORONTO

Copyright © 1974 by John Wiley & Sons. Inc.

All rights reserved. Published simultaneously in Canada.

No part of this book may be reproduced by any means, nor transmitted, nor translated into a machine language without the written permission of the publisher.

"John Wiley & Sons, Inc. is pleased to publish this volume of Organic Syntheses on behalf of Organic Syntheses. Inc. Although Organic Syntheses. Inc. has assured us that each preparation contained in this volume has been checked in an independent laboratory and that any hazards that were uncovered are clearly set forth in the write-up of each preparation, John Wiley & Sons, Inc. does not warrant the preparations against any safety hazards and assumes no liability with respect to the use of the preparations."

Library of Congress Catalog Card Number: 21-17747

ISBN 0-471-42836-1

Printed in the United States of America

10 9 8 7 6 5 4 3 2 1

NOMENCLATURE

Preparations appear in the alphabetical order of common names of the compounds or names of the synthetic procedures. For convenience in surveying the literature concerning any preparation through *Chemical Abstracts* subject indexes, the *Chemical Abstracts* indexing name for each compound is given as a subtitle if it differs from the common name used as the title.

SUBMISSION OF PREPARATIONS

Chemists are invited to submit for publication in *Organic Syntheses* procedures for the preparation of compounds that are of general interest, as well as procedures that illustrate synthetic methods of general utility. It is fundamental to the usefulness of *Organic Syntheses* that submitted procedures represent optimum conditions, and the procedures should have been checked carefully by the submitters, not only for yield and physical properties of the products, but also for any hazards that may be involved. Full details of all manipulations should be described, and the range of yield should be reported rather than the maximum yield obtainable by an operator who has had considerable experience with the preparation. For each solid product the melting-point range should be reported, and for each liquid product the range of boiling point and refractive index should be included. In most instances, it is desirable to include additional physical properties of the product, such as ultraviolet, infrared, mass, or nuclear magnetic resonance spectra, and criteria of purity such as gas chromatographic data. In the event that any of the reactants are not readily commercially available at reasonable cost, their preparation should be described in a complete detail and in the same manner as the preparation of the product of major interest. The sources of the reactants should be described in notes, and the physical properties (such as boiling point, index of refraction, melting point) of the reactants should be included except where standard commercial grades are specified.

Beginning with Volume 49, Sec. 3., Method of Preparation, and Sec. 4., Merits of the Preparation, have been combined into a single new Sec. 3., Discussion. In this section should be described other practical methods for accomplishing the purpose of the procedure that have appeared in the literature. It is unnecessary to mention methods that have been published but are of no practical synthetic value. Those features of the procedure that recommend it for publication in *Organic Syntheses* should be cited (synthetic method of considerable scope, specific compound of interest not likely to be made available commercially, method that give better yield or is less laborious than other methods, etc.). If possible, a brief discussion of the scope and limitations of the procedure as applied to other examples as well as a comparison of the method with the other methods cited should be included. If necessary to the understanding or use of the method for related syntheses, a brief discussion of the mechanism may be placed in this section. The present emphasis of *Organic Syntheses* is on model procedures rather than on specific compounds (although the latter are still welcomed), and the Discussion section should be written to help the reader decide whether and how to use the procedure in his own research. Three copies of each procedure should be submitted to the Secretary of the Editorial Board. It is sometimes helpful to the Board if there is an accompanying letter setting forth the features of the preparations that are of interest.

Additions, corrections, and improvements to the preparations previously published are welcomed and should be directed to the Secretary.

PREFACE

This volume of *Organic Syntheses* represents a compilation of useful methodology and versatile reagents and includes several multistage syntheses of dramatic, if not heroic, proportions. The latter work represents somewhat of a departure from the traditional *Organic Syntheses* style, but serves to emphasize the advances made in synthesis. In addition to these features, this volume includes several procedures that provide a comparison between alternate methods for accomplishing a similar transformation. With no intention of slighting any of the excellent procedures included, several of the forementioned deserve further comment.

Through the concerned efforts of their contributors and the magnanimous contributions of scientific expertise, I am fortunate to be able to include as an *Organic Syntheses* procedure the synthesis of [18]ANNULENE and 1,6-METHANO[10]ANNULENE. The syntheses of both compounds represent landmarks in organic chemistry, and their availability through *Organic Syntheses* procedures may spur further work on the interesting structures. Equally valuable will be the preparation of *trans*-4-HYDROXY-2-HEXENAL via 1,3-bis(methylthio)allyllithium and the procedure for the formation of CYCLOPROPYLDIPHENYLSULFONIUM FLUOROBORATE. These experiments represent further examples of the utility of sulfur-containing reagents in synthesis. Similarly, the formation of 2,2-(TRIMETHYLENEDITHIO)CYCLOHEXANONE and 2,2-(ETHYLENEDITHIO)CYCLOHEXANONE from TRIMETHYLENE DITHIOTOSYLATE and ETHYLENE DITHIOTOSYLATE presents detailed procedures for the use of this valuable blocking group.

Carbonyl alkylation and condensation reactions are always of great value in synthesis, and the formation of *o*-ANISALDEHYDE via 4,4-dimethyl-2-oxazoline, 2,2-DIMETHYL-3-PHENYLPROPIONALDEHYDE via alkylation of the magnesio-enamine salt and *threo*-4-HYDROXY-3-PHENYL-2-HEPTANONE via a directed aldol condensation represent excellent conditions to the synthetic methodology in this area.

Several comparative procedures are included. The formation of 1-BENZYLINDOLE and GERANYL CHLORIDE by two different procedures are representative. An interesting comparison of three of the recent adaptations of the Claisen rearrangement on the same substrate is presented in the preparations of N,N-DIMETHYL-5β-CHOLEST-3-ENE-5-ACETAMIDE, ETHYL-5β-CHOLEST-3-ENE-5-ACETATE, and 5β-CHOLEST-3-ENE-5-ACETALDEHYDE. For the utility of the procedure itself as well as for comparison with previously presented syntheses, the preparation and use of triflates in the synthesis of CYCLOBUTANONE is included.

The preparation of macrocyclic diimines and endocyclic enamines are represented by the procedures for the formation of 1,10-DIAZACYCLOÖCTADECANE and N-METHYL-2-PHENYL-Δ^2-TETRAHYDROPYRIDINE. Other procedures representative of alkylation reactions and aromaticity (TRI-t-BUTYLCYCLOPROPENYL FLUOROBORATE) round out a volume of tested experimental procedures of general value.

The Board of Editors thanks the contributors of the preparations and encourages the continued interest of the scientific community in this project through their suggestions for changes that will improve the usefulness of *Organic Syntheses*. This effort will only remain viable as long as it serves the needs of chemists, and to do that the Board of Editors must rely on everyone's advice and help. Continued active contribution of procedures that represent new techniques and useful new compounds or reagents is solicited at all times by the Board and every effort will be made to publish these procedures as rapidly as possible. Scientific life is always fast and furious, and everyone has many, too many, demands on his time, but the Editors will always try to make the time spent in preparing a contribution for *Organic Syntheses* a worthy effort, and one that the scientific community as a whole will appreciate.

This volume continues the recent "tradition" of including a listing of unchecked preparations that have been received during the preceding year. These are available from the Secretary's office for a nominal fee prior to checking. The Board of Editors would appreciate receiving any comments on this practice, as well as the procedures themselves, from those who avail themselves of this service.

PREFACE

As Editor-in-Chief I have many dedicated people to thank for a richly rewarding experience and for making this volume possible. My colleagues on the Board of Editors and *their* untiring collaborators have considered, cajoled, coddled, and *checked* many of the procedures included in this volume, and without their help there would be no volume. The contributions of my own co-workers must not go unheralded as they, too, took on their tasks with an interest and zeal that was very heartwarming to the "boss." When the science and experimentation are over, a volume of *Organic Syntheses* is far from being born, and the credits must continue at length throughout the publication process. My gratitude is particularly boundless to the midwife in the birth of this manuscript. Mrs. Carla Willard not only typed toward an unrealistic deadline but made it and still retained her bright and reassuring composure. Finally, I want to express my gratitude to Dr. S. Kasparek for preparing the Contents and Author and Subject Indexes—a task that required his talent and for which I was ill-equipped.

Pasadena, California   ROBERT E. IRELAND
June 1974

CONTENTS

[18]ANNULENE	1
1,6-METHANO[10]ANNULENE	11
γ-HYDROXY-α,β-UNSATURATED ALDEHYDES via 1,3-BIS(METHYLTHIO) ALLYLLITHIUM: trans-4-HYDROXY-2-HEXENAL	19
CYCLOPROPYLDIPHENYLSULFONIUM FLUOROBORATE	27
TRIMETHYLENE DITHIOTOSYLATE AND ETHYLENE DITHIOTOSYLATE	33
2,2-(ETHYLENEDITHIO)CYCLOHEXANONE	37
2,2-(TRIMETHYLENEDITHIO)CYCLOHEXANONE	39
ALDEHYDES FROM 4,4-DIMETHYL-2-OXAZOLINE AND GRIGNARD REAGENTS: o-ANISALDEHYDE	42
ALKYLATIONS OF ALDEHYDES via REACTION OF THE MAGNESIOENAMINE SALT OF AN ALDEHYDE: 2,2-DIMETHYL-3-PHENYLPROPIONALDEHYDE	46
DIRECTED ALDOL CONDENSATIONS: threo-4-HYDROXY-3-PHENYL-2-HEPTANONE	49
1-BENZYLINDOLE	58
N-ALKYLINDOLES FROM THE ALKYLATION OF INDOLE SODIUM IN HEXAMETHYLPHOSPHORAMIDE: 1-BENZYLINDOLE	60
GERANYL CHLORIDE	63
ALLYLIC CHLORIDES FROM ALLYLIC ALCOHOLS: GERANYL CHLORIDE	68
5-β CHOLEST-3-ENE-5-ACETALDEHYDE	71
ETHYL 5β-CHOLEST-3-ENE-5-ACETATE	74
N,N-DIMETHYL-5β-CHOLEST-3-ENE-5-ACETAMIDE	77
PREPARATION OF VINYL TRIFLUOROMETHANESULFONATES: 3-METHYL-2-BUTEN-2-YL-TRIFLATE	79
CYCLOBUTANONE via SOLVOLYTIC CYCLIZATION	84
MACROCYCLIC DIIMINES: 1,10-DIAZACYCLOÖCTADECANE	88
ENDOCYCLIC ENAMINE SYNTHESIS: N-METHYL-2-PHENYL-Δ^2-TETRAHYDROPYRIDINE	93
TRI-tert-BUTYLCYCLOPROPENYL FLUOROBORATE	97
CUMULATIVE AUTHOR INDEX, VOLUMES 50 TO 54	103
CUMULATIVE SUBJECT INDEX, VOLUMES 50 TO 54	107
UNCHECKED PROCEDURES	133

[18]ANNULENE

(1,3,5,7,9,11,13,15,17-cycloöctadecanonaene)

A.

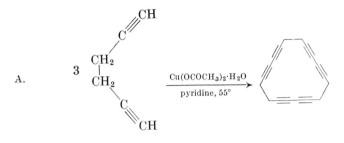

B.

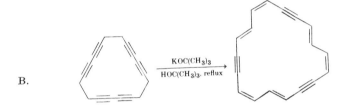

C.

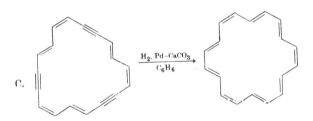

Submitted by K. STÖCKEL and F. SONDHEIMER[1]
Checked by R. C. WHELAND, R. E. BENSON, H. ONA, and S. MASAMUNE

1. Procedure

A. *Oxidative coupling of 1,5-hexadiyne.* A 12-l., four-necked, round-bottomed flask provided with a stopcock at the bottom (Figure 1) (Note 1) is fitted with a stirrer (Note 2), a reflux condenser, and a 500-ml. dropping funnel equipped with a pressure-equalizing side arm. The

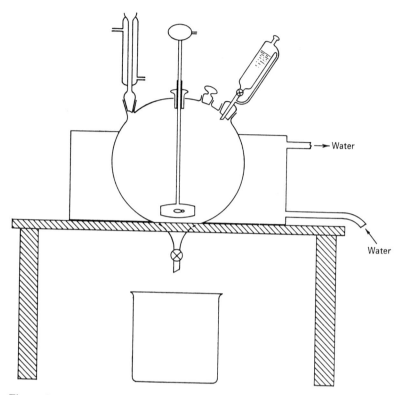

Figure 1.

flask is placed in a large metal vessel equipped with a hot–cold water inlet and a drain. The flask is charged with 600 g. (3 moles) of cupric acetate monohydrate (Note 3) and 3.8 l. of pyridine (Note 4). Stirring is begun and warm water is added to the metal vessel to heat the contents of the flask to 55 ± 1°, and the mixture is slowly stirred at this temperature for 1 hour. A solution of 50 g. (0.64 mole) of 1,5-hexadiyne (Note 5) in 400 ml. of pyridine (Note 4) is then added during 30 minutes to the vigorously stirred (approximately 600 r.p.m.) green suspension at 55° (Note 6), and the mixture is stirred vigorously at this temperature for an additional 2 hours. The warm water in the metal vessel is allowed to drain and is replaced with a mixture of ice and water. When the contents of the flask have cooled, the mixture is drawn off through the stopcock at the bottom of the flask and filtered through a 25-cm. Büchner funnel (Filtrate A). The green residue is transferred to a 5-l.

beaker, 2.5 l. of a benzene–ether mixture (1:1) is added, and the resulting mixture is stirred well to achieve good mixing without emulsion formation. The resulting mixture is filtered (Filtrate B), and the residue is similarly extracted with 2.5 l. of a benzene–ether mixture (1:1) and filtered (Filtrate C). Each filtrate is kept separate.

Filtrate A is concentrated to approximately 200 ml. (Note 7), and the resulting slurry is added to Filtrate C. The resulting mixture is well agitated and then filtered through a 25-cm. Büchner funnel. The residue is extracted twice in a similar manner with 500 ml. of a benzene–ether mixture (1:1). All the filtrates are now combined in the original 12-l. reaction vessel (Note 8) and washed successively with two 500-ml. portions of water, two 500-ml. portions of cold (0°) $2N$ hydrochloric acid, and three 500-ml. portions of water. The organic layer is dried for 2 hours over about 110 g. of magnesium sulfate (Note 9). After the drying agent is removed by filtration, evaporation of the filtrate to dryness (Note 7) affords approximately 16 g. of crude product.

B. *Tridehydro[18]annulene.* The dark brown residue from Part A above and 800 ml. of benzene (Note 10) are added to a 2-l., round-bottomed flask fitted with a reflux condenser and a calcium chloride-containing drying tube. The solid dissolves on heating to reflux, and then the solution is allowed to cool briefly. The condenser is removed, and 800 ml. of a saturated solution of potassium t-butoxide in t-butyl alcohol (Notes 11 and 12) is rapidly added (Note 13). The resulting solution is heated to reflux for 30 minutes. The hot mixture is transferred to the original 12-l. reaction vessel (Note 14), containing approximately 2 l. of ice, and 1.5 l. of ether is added. The resulting mixture is well stirred and then allowed to stand for a few minutes. The incompletely separated phases, in which is suspended a large amount of black solid, are filtered through a 25-cm. Büchner funnel. Whenever the filtration becomes slow, the black material on the filter paper is washed with ether, and the paper is replaced. The organic layer of the filtrate is separated, washed three times with 500-ml. portions of water, and dried for 2 hours over approximately 100 g. of anhydrous magnesium sulfate (Note 9). The drying agent is removed by filtration, and the filtrate is evaporated to dryness (Notes 7 and 15). The red-brown viscous residue is dissolved in 60 ml. of benzene (Note 10) and chromatographed on alumina.

One liter of a pentane–ether mixture (95:5) (Note 16) is poured into a closed chromatography column (100 × 4.5 cm.), the bottom of which is protected by a small plug of cotton wool. One kilogram of alumina (Note 17) is added in portions, with slow rotation by hand. The stopcock is opened, and the level of the supernatant liquid is allowed to fall to the top of the alumina. The benzene solution is now added with a long pipette and is allowed to seep in. The column is developed first with 1 l. of pentane–ether (95:5) in order to wash out the benzene and then with pentane–ether (80:20). Two bands are observable on the column. The faster moving, light brown band consists of tridehydro[18]annulene (Note 18) and the slower moving, dark brown band consists of tetradehydro[24]annulene. When the first band is approximately 15 cm. from the bottom of the column, after approximately 4–6 l. of solvent have been eluted, 150-ml. fractions are collected, and the electronic spectrum of each fraction is determined. As soon as the maxima at 385 and 400 nm. characteristic of tridehydro[18]annulene appear (Note 19), the fractions are combined ("mixed fractions") until the first band is approximately 1 cm. from the bottom of the column. Essentially "pure" tridehydro[18]annulene (Note 18) is now eluted, and this material is collected until the second band is approximately 5 cm. from the bottom of the column. Smaller fractions (150 ml. each) are now collected again; the electronic spectrum of each fraction is determined, and those still showing maxima at 385 and 400 nm. are combined with the previously mentioned "mixed fractions." If required, the tetradehydro[24]annulene suitable for conversion to [24]annulene[2] can then be eluted with pentane–ether (70:30 to 50:50).

The electronic spectrum of the fractions containing the "pure" tridehydro[18]annulene exhibits the strongest absorption maximum (in benzene) at 342 nm. (ϵ 155,000)[3] and the spectroscopic yield, based on the molar extinction coefficient, is 1.17 g. (2.40% from 1,5-hexadiyne). The yield of tridehydro[18]annulene in the "mixed fractions," based on the 342 nm. maximum, is 0.27 g. (0.55%). The tridehydro[18]annulene is best stored in solution in the refrigerator.

C. *[18]Annulene.* In a 300-ml. conical flask fitted with a side arm (with a closed stopcock) and a 3.5-cm. Teflon®-coated magnetic stirring bar is placed 1 g. of a 10% palladium–calcium carbonate catalyst (Note 20) and 30 ml. of benzene (Note 10). The flask is attached to an atmospheric pressure hydrogenation apparatus, and the air in the system is replaced by hydrogen by evacuating the flask and refilling with hydrogen three times. The catalyst suspension is stirred until no more gas is absorbed. One-third (390 mg., determined spectroscopically) of the "pure" tridehydro[18]annulene solution described in Part B is evaporated to dryness (Note 7), and the resulting brown crystalline residue is dissolved in 30 ml. of benzene (Note 10). This solution is added to the hydrogenation flask through the side arm without stirring. The mixture is now stirred under a hydrogen atmosphere as rapidly as possible (*ca.* 600 r.p.m.) (Note 21) until 207 ml. (4.9 molar equivalents) of gas (22°, 740 mm.) are absorbed *ca.* 5 minutes. The mixture is filtered through a sintered glass funnel, and the catalyst is washed with three 20-ml. portions of benzene.

The above hydrogenation experiment is repeated twice with the remaining two-thirds of the "pure" tridehydro[18]annulene solution, and the combined filtrates from the three hydrogenations are evaporated to approximately 30 ml. (Note 7). The solution is then transferred with a pipette to a test tube (10 × 3 cm.) and concentrated to approximately 15 ml. at 40° by means of a stream of nitrogen. The dark green solution is diluted with 20 ml. of ether and cooled in an ice bath for 30 minutes. The resulting red-brown crystals of [18]annulene are collected by filtration on a sintered glass funnel and are washed with approximately 3 ml. of cold ($-20°$) ether. After drying in air for a few minutes this material weighs 114 mg. A second crop that amounts to 42 mg. is obtained by evaporation of the mother liquors to dryness (Note 7), followed by crystallization from 6 ml. of benzene and 20 ml. of ether. Both crops give a single spot on thin-layer chromatography (Note 22).

The [18]annulene mother liquors contain 181 mg. of unchanged tridehydro[18]annulene, as determined by the electronic spectrum (see Part B). They are combined with the "mixed fractions" described in Part B (containing 270 mg. of tridehydro[18]annulene), and the resulting solution is evaporated to dryness (Note 7). The residue is dissolved in 30 ml. of benzene. This solution is stirred with hydrogen and 1.0 g. of a 10% palladium–calcium carbonate catalyst (Note 20), as before, until 216 ml. (4.4 molar equivalents) of hydrogen (20°, 740 mm.) are absorbed. The catalyst is removed, and the filtrate is concentrated to approximately 6 ml., as described previously. After the addition of 20 ml. of ether and cooling in an ice bath, the resulting crystals are collected by filtration and amount to a further 112 mg. of [18]annulene (Note 23). The purity of the product is confirmed by thin layer chromatography (Note 22). The combined yield of [18]annulene is 268 mg. (0.54% overall from 1,5-hexadiyne) (Note 24).

[18]Annulene is best stored in benzene–ether solution in the refrigerator. The crystals can also be kept for some time in the refrigerator, small crystals being less stable than large ones.[2] Material that has decomposed to some extent may be purified as follows. The substance (300 mg.) is broken up with a spatula and heated with 30 ml. of benzene, and the insoluble material is removed by filtration. The filtrate is poured onto a chromatography column (20 × 1.8 cm.), prepared with 35 g. of alumina (Note 17) and benzene. The column is washed with benzene until the eluate is colorless. Polymeric material is retained at the top of the column. The eluent is concentrated to approximately 6 ml., as previously, diluted with 20 ml. of ether, and cooled in an ice–salt bath. The resulting crystals of [18]annulene are then collected on a sintered glass funnel, washed with 3 ml. of cold (−20°) ether, and dried in air for a few minutes.

2. Notes

1. The submitters used an ordinary 10-l., three-necked, round-bottomed flask.

2. A stirrer with one 11-cm., Teflon® blade was employed by the checkers. The submitters used a stirrer with two 7.5-cm. paddles, 9 cm. apart.

3. Cupric acetate monohydrate A. R. available from Fisons Scientific Apparatus Ltd. (Loughborough, England) was employed by the submitters. The checkers used both Baker Analytical Reagent and Fisher A.C.S. certified cupric acetate monohydrate. Small portions were finely ground with a mortar and pestle.

4. The submitters employed pyridine from BDH Chemicals Ltd. (Poole, England) that was dried over solid potassium hydroxide for 24 hours and then distilled under slightly reduced pressure. The checkers used a similarly treated product.

5. 1,5-Hexadiyne was obtained from Farchan Research Laboratories and was distilled at atmospheric pressure (b.p. 84–85°) before use. It can be prepared as described by Raphael and Sondheimer.[4]

6. Vigorous stirring is important, since the yield of product appears to be reduced if the mixture is not stirred well (J. E. Fox, unpublished observation). The checkers used rates of 530–560 r.p.m.

7. All evaporations, unless otherwise stated, were carried out with a rotary evaporator under reduced pressure (water pump) in a water bath kept at about 40°. The checkers' water bath was maintained at

44–50°, and the evaporation required 4–5 hours rather than the 2 hours found by the submitters.

8. The submitters used a 10-l. separatory funnel.

9. Magnesium sulfate ("dried") from BDH Chemicals Ltd. (Poole, England) was used by the submitters. The checkers used anhydrous magnesium sulfate available from Fisher Scientific Company.

10. Benzene A. R. from Fisons Scientific Apparatus Ltd. (Loughborough, England) was employed by the submitters, and Fisher reagent grade benzene was used by the checkers.

11. The solution was obtained from 44 g. of potassium (Fisons Scientific Apparatus Ltd., Loughborough, England, submitters; Fisher Scientific Company, checkers) and 1 l. of t-butyl alcohol (Note 12) by boiling under reflux under nitrogen in the apparatus described by Vogel[5] until all the metal had dissolved. The solution is kept well stoppered. If crystals of potassium t-butoxide separate, they are dissolved by gentle heating before use.

12. t-Butyl alcohol from May & Baker Ltd. of Dagenham, England, was dried by refluxing 1 l. of the alcohol with approximately 40 g. of sodium wire until about two-thirds of the metal had dissolved, and, then recovering the alcohol by distillation (see Note 2 in Reference 5). The checkers used t-butyl alcohol available from Fisher Scientific Company and recommend conducting this purification on twofold scale.

13. An exothermic reaction was observed at this point. Too rapid addition of the t-butoxide solution to the benzene solution will lead to loss of product by frothing.

14. The submitters used a 10-l. separatory funnel and shook the mixture after addition to the ice and ether.

15. It is important that all volatile solvents are removed, particularly t-butyl alcohol, since its presence will interfere with the subsequent chromatography procedure. The checkers obtained approximately 9.0 g. of crude material.

16. The submitters used "pentane" petroleum spirit (b.p. 25–40°) available from Hopkin & Williams Ltd., Hadwell Heath, England, and anhydrous diethyl ether available from May & Baker Ltd., Dagenham, England. Both solvents were each distilled over solid potassium hydroxide before use. The checkers purchased pentane from Phillips Petroleum Company and distilled it over solid potassium hydroxide. Anhydrous diethyl ether supplied by Mallinckrodt Chemical Works was used without further purification.

17. Aluminium oxide (neutral, activity grade I) available from M. Woelm, Eschwege, Germany, was deactivated by the addition of 7 ml. of water to 1 kg. of the adsorbent before use.

18. "Tridehydro[18]annulene" here and in the sequel refers to the symmetrical isomer shown in the formula, admixed with smaller quantities of an unsymmetrical isomer and tetradehydro[18]annulene.[6] These can be separated by chromatography on alumina coated with 20% silver nitrate, but this is unnecessary for the synthesis of [18]annulene since all three substances give this annulene on catalytic hydrogenation.[6]

19. For the full electronic spectra of tridehydro[18]annulene and tetradehydro[24]annulene in isoöctane, see Reference 3.

20. The submitters used 10% palladium–calcium carbonate available from Fluka AG, Buchs, Switzerland. The checkers prepared the catalyst in a manner similar to that described by M. Busch and H. Stöve.[7] To a solution of calcium chloride (5.1 g.) in 60 ml. of water was added 4.8 g. of sodium carbonate. The precipitate was filtered, washed thoroughly with water, and then suspended in 30 ml. of water. To this calcium carbonate suspension was added with stirring a solution containing 0.833 g. of palladium chloride and a few drops of $6N$ hydrochloric acid. The catalyst mixture was then filtered through a sintered glass funnel, and the solid was washed with water until the chloride test (silver nitrate) became negative. After being washed with ethanol and ether, the catalyst was dried under reduced pressure.

21. Vigorous stirring is essential, since the yield of [18]annulene is reduced considerably if the mixture is stirred more slowly (R. Wolovsky, unpublished observation).

22. Thin-layer chromatography was performed on silica gel GF_{254} plates supplied by E. Merck AG, Darmstadt, Germany, using the solvent system pentane–cyclohexane–benzene (92:4:4). The electronic spectra in benzene were essentially identical to those previously reported,[2] and exhibit absorption maxima at 378 (ϵ 270,000), 414 (ϵ 8,600), 428 (6,700), and 455 nm. (26,200). Proton magnetic resonance spectra of [18]annulene (deuteriotetrahydrofuran solution, vacuum-sealed, tetramethylsilane reference) are temperature-dependent and show a singlet at 5.45 p.p.m. at 120° and two multiplets at 9.25 ($12H$) and -2.9 p.p.m. ($6H$) at $-60°$. The latter two signals merge just above room temperature.[8] The checkers observed the similar temperature dependence of ^{13}C magnetic resonance spectra (deuteriotetrahydrofuran, tetramethylsilane reference, 22.6 MHz) of the [18]annulene. Thus,

proton-decoupled spectra show a singlet at 126 p.p.m. at 60° and two singlets at 128 and 121 p.p.m. at −70°. The rapid processes exchanging the inner and outer nuclei (both proton and ^{13}C) in solution are responsible for the above spectral behavior.

23. The submitters reported in their original procedure that some additional amounts of [18]annulene and tridehydro[18]annulene were obtained by chromatography of the mother liquors.

24. The yields described in Parts B and C are the average values of two runs performed by the checkers, and both runs proceeded with virtually identical results at every stage. The submitters obtained a 0.63% overall yield of [18]annulene from 1,5-hexadiyne.

3. Discussion

Only two general methods have been developed for the synthesis of the macrocyclic annulenes.[9] The first of these, developed by Sondheimer and co-workers, involves the oxidative coupling of a suitable terminal diacetylene to a macrocyclic polyacetylene of required ring size, using typically cupric acetate in pyridine. The cyclic compound is then transformed to a dehydroannulene, usually by prototropic rearrangement effected by potassium *t*-butoxide. Finally, partial catalytic hydrogenation of the triple bonds to double bonds leads to the annulene.

The presently described procedure for the synthesis of [18]annulene, although the overall yield is low by the standard normally set for *Organic Syntheses*, illustrates the above general route leading to the theoretically important macrocyclic annulenes, and in this way [14]-, [16]-, [18]-, [20]-, [22]-, and [24]annulenes have been prepared in pure crystalline form.

[18]Annulene was the first macrocyclic annulene containing $(4n + 2)$ π-electrons to be synthesized. The compound is of considerable interest, since it is the type of annulene that was predicted to be aromatic by Hückel.[10] It proved to be aromatic in practice, as evidenced from the proton magnetic resonance spectrum,[8,11] the X-ray crystallographic analysis,[12] and the fact that electrophilic substitution reactions could be effected.[13]

The method of synthesis is essentially that described by Sondheimer and Wolovsky[3] (preparation of tridehydro[18]annulene) and by Sondheimer, Wolovsky, and Amiel[2] (hydrogenation of tridehydro[18]-annulene to [18]annulene). It has been simplified, since [18]annulene is now obtained from tridehydro[18]annulene without chromatography, and the whole procedure involves only one chromatographic separation. [18]Annulene has also been obtained by Figeys and Gelbcke[14] in 0.42% overall yield by a six-step sequence from propargyl alcohol via propargyl aldehyde, *meso*-1,5-hexadiyne-3,4-diol, *meso*-1,5-hexadiyne-3,4-diol ditosylate, *cis*-3-hexene-1,5-diyne, and 1,3,7,9,13,15-hexadehydro[18]annulene.

1. Chemistry Department, University College, London WC1H OAJ, England.
2. F. Sondheimer, R. Wolovsky, and Y. Amiel, *J. Amer. Chem. Soc.*, **84**, 274 (1962).
3. F. Sondheimer and R. Wolovsky, *J. Amer. Chem. Soc.*, **84**, 260 (1962).
4. R. A. Raphael and F. Sondheimer, *J. Chem. Soc.*, 120 (1950).
5. A. I. Vogel, "Practical Organic Chemistry," 3rd ed., Longmans, Green and Co., London, 1967, p. 921.
6. R. Wolovsky, *J. Amer. Chem. Soc.*, **87**, 3638 (1965).
7. M. Busch and H. Stöve, *Chem. Ber.*, **49**, 1063 (1916).
8. J.-M. Gilles, J. F. M. Oth, F. Sondheimer, and E. P. Woo, *J. Chem. Soc. B.*, 2177 (1971).
9. For a review of the annulenes, see F. Sondheimer, *Accounts Chem. Res.*, **5**, 81 (1972).
10. E. Hückel, *Z. Physik*, **70**, 204 (1931); "Grundzüge der Theorie Ungesättigter und Aromatischer Verbindungen," Verlag Chemie, Berlin, 1938.
11. L. M. Jackman, F. Sondheimer, Y. Amiel, D. A. Ben-Efraim, Y. Gaoni, R. Wolovsky, and A. A. Bothner-By, *J. Amer. Chem. Soc.*, **84**, 4307 (1962).
12. J. Bregman, F. L. Hirschfeld, D. Rabinovich and G. M. J. Schmidt, *Acta Crystallogr.* **19**, 227 (1965); F. L. Hirschfeld and D. Rabinovich, *Acta Crystallogr.*, **19**, 235 (1965).
13. I. C. Calder, P. J. Garratt, H. C. Longuet-Higgins, F. Sondheimer, and R. Wolovsky, *J. Chem. Soc. C*, 1041 (1967); E. P. Woo and F. Sondheimer, *Tetrahedron*, **26**, 3933 (1970).
14. H. P. Figeys and M. Gelbcke, *Tetrahedron Lett.*, 5139 (1970).

1,6-METHANO[10]ANNULENE

A. [naphthalene] —Na/NH₃, C₂H₅OH→ [1,4,5,8-tetrahydronaphthalene]

B. [isotetralin] —CHCl₃, (CH₃)₃COK→ [dichlorocarbene adduct]

C. [dichloro adduct] —Na/NH₃→ [methano compound]

D. [methano dihydro compound] —DDQ, Dioxane→ [1,6-methano[10]annulene]

Submitted by E. VOGEL,[1] W. KLUG, and A. BREUER
Checked by R. E. IRELAND, R. A. FARR, H. A. KIRST, T. C. MCKENZIE, R. H. MUELLER, R. R. SCHMIDT, III, D. M. WALBA, A. K. WILLARD, and S. R. WILSON

1. Procedure

Caution! This reaction should be carried out in an efficient hood.

A. *1,4,5,8-Tetrahydronaphthalene (Isotetralin).* In a 12-l. (Note 1), three-necked, round-bottomed flask, immersed into a dry ice–acetone bath and fitted with a dry ice condenser (Note 2), a tube-sealed stirrer (Note 3), a drying tube (potassium hydroxide), and a gas delivery-tube running to the bottom of the flask, 3 l. of ammonia is condensed (Note 4). The gas delivery tube is removed, and with vigorous stirring, 192.3 g. (8.4 g.-atoms) of sodium is added in small portions (Note 5) during a period of 1 hour. The flask is then fitted with a dropping funnel, through which a solution of 192.3 g. (1.5 mole) of naphthalene in a mixture of 750 ml. of ether and 600 ml. of ethanol is added dropwise to the blue solution during 3 hours. After the addition is complete, the reaction mixture is stirred at $-78°$ (Note 6) for another 6 hours. The cooling bath is removed, and the ammonia is allowed to evaporate overnight. The remaining white solid residue is processed with ice cooling and stirring under a nitrogen atmosphere by slow addition of first 120 ml. of methanol to destroy unreacted sodium and then 4–5 l. of ice water to dissolve the salts (Note 7). The reaction mixture is extracted with 1 l. of ether. Evaporation of the ether phase at room temperature under reduced pressure gives a coarse, white solid which is collected on a sintered-glass funnel and washed with water. Recrystallization from methanol (about 1.6 l.) using a heated funnel followed by drying of the crystals under reduced pressure (Note 8) gives 148–158 g. (75–80%) of isotetralin, m.p. 52–53° (purity $\sim$98%) (Note 9). Pure 1,4,5,8-tetrahydronaphthalene is reported to have m.p. 58°.[2]

B. *11,11-Dichlorotricyclo[4.4.1.0^{1,6}]undeca-3,8-diene.* A 3-l., three-necked, round-bottomed flask is fitted with a tube-sealed stirrer, a pressure-equalizing dropping funnel, and a Claisen-adapter that bears an inlet tube for argon and a thermometer for low temperatures. In the flask is dissolved 132.2 g. (1.0 mole) of 1,4,5,8-tetrahydronaphthalene (isotetralin) in 1.3 l. of anhydrous ether (Note 10). To this solution is added 150 g. (1.33 mole) of potassium *t*-butoxide (Note 11) under an argon atmosphere, and the resulting suspension is cooled to $-30°$ with a dry ice–acetone bath and stirred efficiently. While these conditions are maintained, a solution of 119.5 g. (1.0 mole) of chloroform in 150 ml. of ether is added dropwise during 90 minutes (Note 12). The mixture is stirred for another 30 minutes at $-30°$, and the temperature is then allowed to rise above 0°. Following this, 300–350 ml. of ice water is added to dissolve the salts. The two layers which form are separated (Note 13). The organic layer is washed with two 300-ml. portions of water while the aqueous layer is extracted with two 200-ml. portions of ether. The ether phases are combined and dried over magnesium sulfate. After filtration from the drying agent, the ether is removed on a rotary evaporator, and the residual liquid (or solid) is distilled under reduced pressure.

The distillation is expediently carried out in a set-up which consists of a 500-ml., round-bottomed flask, an electrically heated 1.5 × 30 cm. column packed with V4A wire spirals (4 mm.) (Note 14), a short, air-cooled condenser (Note 15), and an ice-cooled, three-necked, 250-ml.

receiver flask. During the distillation, the liquid is stirred by a magnetic stirring bar and heated by an oil bath by means of a stirrer hot plate.

The first fraction, b.p. 55–58° (1 mm.) is isotetralin (~50 g.) (Note 16). Some more of this material distills when the column is heated to about 100°. The temperature in the head of the column thereby rises to 90–95°, and it is then necessary to change the receiver flask. The second fraction, b.p. 95–102° (1 mm.) constitutes the 1:1-adducts of dichlorocarbene to isotetralin (~108 g.), which consist of 92% of 11,11-dichlorotricyclo[4.4.1.0^{1,6}]undeca-3,8-diene and 8% of the side-addition product. The residue mainly contains the 2:1-adducts. The fraction of the 1:1-adducts is recrystallized from methanol (about 500 ml.) to give 87–97 g. (40–45%, based on isotetralin initially used) of 11,11-dichlorotricyclo[4.4.1.0^{1,6}]undeca-3,8-diene as long colorless needles, m.p. 88–89° (Note 17).

C. *Tricyclo[4.4.1.0^{1,6}]undeca-3,8-diene.* In a 2-l., three-necked, round-bottomed flask, immersed into a dry ice–acetone bath and fitted with a dry ice condenser, a tube-sealed stirrer, a drying tube, and a gas delivery tube running to the bottom of the flask, 800 ml. of ammonia is condensed. The gas delivery tube is removed and with vigorous stirring 56 g. (2.45 g.-atoms) of sodium is added in small portions over a period ½ hour (Note 18). The flask is then fitted with a dropping funnel through which a solution of 81.4 g. (0.378 mole) of 11,11-dichlorotricyclo[4.4.1.0^{1,6}]undeca-3,8-diene (m.p. 88–89°) in 500 ml. of anhydrous ether is added during an hour, while cooling and stirring is maintained. After the addition is complete, the dry ice–acetone bath is removed, and the ammonia is allowed to evaporate overnight. The flask is placed into the dry ice–acetone bath again, and a gentle stream of argon is passed continuously through the system. With stirring, a mixture of 90 ml. of methanol and 90 ml. of ether is then added drop wise. Following this, the bath temperature is allowed to rise to 0° and, with continued stirring, 500 ml. of ice water is added slowly. The reaction mixture is transferred to a 2-l. separatory funnel, and the two layers are separated. The organic layer is washed with 200 ml. of water, while the aqueous layer is extracted with three 150-ml. portions of pentane (Note 19). The combined ether–pentane phases are dried over magnesium sulfate. After filtration of the drying agent (Note 20) the solvent is removed by distillation through a 30-cm. Vigreux column. The liquid that remains is transferred to a 250-ml., round-bottomed flask and distilled under reduced pressure through a packed column

(Note 21). Tricyclo[4.4.1.0^{1,6}]undeca-3,8-diene is collected as a colorless liquid at 80–81° (11 mm.) and weighs 46.9–49.7 g. (85–90%), n^{20}D = 1.5180 (Note 22).

D. *1,6-Methano[10]annulene.* A 2-l., three-necked, round-bottomed flask, fitted with a tube-sealed stirrer, a reflux condenser protected with a calcium chloride drying tube and an inlet tube for argon is charged with 900 ml. of anhydrous dioxane (Note 23). To this solvent is added with stirring 149 g. (0.66 mole) of 2,3-dichloro-5,6-dicyano-1,4-benzoquinone (DDQ) (Note 24). When the DDQ has dissolved, 43.8 g. (0.30 mole) of tricyclo[4.4.1.0^{1,6}]undeca-3,8-diene and 10 ml. of glacial acetic acid are added. The system is then flushed with argon, and the stirred mixture is heated at reflux for 5 hours. The reaction starts within a few minutes, as evidenced by effervescing of the solution and massive precipitation of the hydroquinone. At the same time the originally red-brown color of the mixture turns almost black. Following the reflux period, the bulk of the dioxane (600–650 ml.) is removed by distillation through a 15-cm. Vigreux column while stirring is maintained. The mixture, which has become pasty, is cooled, and 150 ml. of *n*-hexane is added. The solid is suction filtered, washed on the filter with 500 ml. of warm *n*-hexane, and dried at 100° to give approximately 144 g. (95%) of pure 2,3-dichloro-5,6-dicyanohydroquinone (Note 25). The filtrate and washings are combined and passed through a 5 × 30 cm. column of neutral alumina (Note 26). The column is eluted with *n*-hexane (Note 27). The solvent is removed by distillation through a 30-cm. Vigreux column. The residual liquid is distilled from a 250-ml., round-bottomed flask through a packed column (Note 28) to give faintly yellow 1,6-methano[10]annulene, b.p. 68–72° (1 mm.), yield 36.2–37.0 g. (85–87%). The compound may crystallize in the receiver-flask, m.p. 27–28° (Note 29).

2. Notes

1. It is advisable to use a 10- or 12-l. flask for runs on this scale because the reaction mixture may effervesce if the naphthalene solution is added too quickly.

2. It is necessary to use a dry ice condenser in order to shorten the time required to condense the ammonia (4 hours compared with 6 hours without the condenser). The ammonia tank was warmed with an

air gun during the distillation. The condenser was removed after the ammonia was collected.

3. It is necessary to use a strong motor for stirring since the reaction mixture becomes temporarily rather viscous.

4. One should not pour the liquified ammonia directly out of the cylinder since particles of iron compounds might be carried along, catalyzing the formation of sodium amide. For the exclusion of moisture it is also necessary to use a drying tower (potassium hydroxide) between the cylinder and the flask.

5. The sodium should be cut into small particles to increase the speed of dissolution and to diminish the danger of stirrer blockage by larger pieces.

6. During this period the reaction mixture might turn white. In this case, another portion of sodium must be added until the solution becomes blue again.

7. The white residue should be worked up as soon as possible. On standing the residue gradually turns brown-red due to the formation of decomposition products; isolation of isotetralin then gets difficult, and the yield may drop sharply. The submitters evaporated any remaining ether from the reaction flask at reduced pressure and filtered the water slurry of isotetralin to give the same yield after crystallization.

8. Isotetralin should not be kept in the vacuum of an oil pump for drying longer than is necessary since the compound has a relatively high vapor pressure.

9. A second extraction of the aqueous phase with ether yields an additional 1.5 g. of material. A second crop of isotetralin could be obtained from the mother liquors of the recrystallization, 29.4 g. (m.p 49–52°).

10. All solvents used should be anhydrous.

11. The yield of 11,11-dichlorotricyclo[4.4.1.0^{1,6}]undeca-3,8-diene strongly depends on the quality of the potassium t-butoxide used. Commercially available, sublimed potassium t-butoxide was employed. When freshly sublimed potassium t-butoxide is utilized, yields of up to 45% of 11,11-dichlorotricyclo[4.4.1.0^{1,6}]undeca-3,8-diene can be obtained. Potassium t-butoxide, prepared by the method of Doering,[3] gave yields comparable to those achieved with the commercial product.

12. The stated reaction temperature should be maintained carefully.

Raising the temperature above −30° noticeably reduces the stereoselectivity of the addition of dichlorocarbene to the central double bond of isotetralin, whereas lowering the temperature causes the yield of 1:1-adducts to drop due to partial crystallization of isotetralin.

13. The formation of emulsions may render it difficult to discern the two rather dark layers. In this case it is helpful to acidify with some dilute sulfuric acid.

14. The checkers used an electrically heated, 1.5 × 30 cm. Vigreux column with the same results.

15. To prevent isotetralin and the 1:1-adducts from solidifying in the condenser external heating with an infrared lamp was applied.

16. The recovered isotetralin can be reused.

17. The product is approximately 99% pure by g.l.c. (SE 30 on kieselguhr, 150°). After two or three recrystallizations from methanol, 11,11-dichlorotricyclo[4.4.1.0^{1,6}]undeca-3,8-diene shows m.p. 90–91°.

18. For the preparation of the solution of sodium in liquid ammonia, compare part A.

19. If emulsions occur, it is advisable to acidify with some dilute sulfuric acid to attain a good separation of the two layers.

20. The drying agent should be washed well with pentane.

21. The column used for this distillation is described in part B.

22. Tricyclo[4.4.1.0^{1,6}]undeca-3,8-diene was shown to be approximately 99% pure by g.l.c. (SE-30 on kieselguhr, 150°).

23. The use of anhydrous solvents is necessary to avoid hydrolytic decomposition of 2,3-dichloro-5,6-dicyano-1,4-benzoquinone.

24. Commercially available 2,3-dichloro-5,6-dicyano-1,4-benzoquinone was employed. 1,6-Methano[10]annulene was obtained in equally good yields, when 2,3-dichloro-5,6-dicyano-1,4-benzoquinone, prepared by the method of Walker and Waugh,[4] was utilized.

25. 2,3-Dichloro-5,6-dicyano-1,4-benzoquinone is readily regenerated in good yield from the hydroquinone by oxidation with nitric acid.[4]

26. Brockmann alumina, activity grade II-III, M. Woelm, 344 Eschwege, West Germany.

27. By contrast to the filtrate and washings, which are rather dark, the eluate is yellow due to the color of 1,6-methano[10]annulene.

28. The column used for this distillation is described in part B.

29. The purity of the 1,6-methano[10]annulene was shown by g.l.c. (SE-30 on kieselguhr, 150°) to be higher than 99%. Recrystallization of the hydrocarbon from methanol raises its melting point to 28–29°.

3. Discussion

The procedure described for the Birch reduction of naphthalene is a modification of the methods previously developed by Birch,[5] Hückel,[2] and Grob.[6] Apart from this reduction, no other practical approaches to isotetralin have become known. The scale employed in the present procedure is not mandatory to achieve optimum yields. Equally good yields were realized when the runs were halved or enlarged up to fourfold. In the latter case, however, the apparatus already reaches pilot plant dimensions.

Tricyclo[4.4.1.0^{1,6}]undeca-3,8-diene, the strategic intermediate, in the synthesis of 1,6-methano[10]annulene from naphthalene, can alternatively be obtained in one step by the reaction of isotetralin with the Simmons–Smith reagent.[7] The two-step preparation of tricyclo-[4.4.1.0^{1,6}]undeca-3,8-diene, which is utilized here, has the following merits: (1) dichlorocarbene adds to the central double bond of isotetralin with exceptionally high stereoselectivity (as compared to that of methylene transfer reagents) to give 11,11-dichlorotricyclo[4.4.1.0^{1,6}]-undeca-3,8-diene as a readily isolable, crystalline compound, (2) the transformation of the dichloro compound into tricyclo[4.4.1.0^{1,6}]undeca-3,8-diene by means of sodium in liquid ammonia[8] is a simple operation and affords the product in high yield and purity. The dichlorocarbene employed in the two-step cyclopropanation of isotetralin was generated by the original method of Doering and Hoffmann.[3] Other sources of dichlorocarbene, notably the methods of Parham and Schweizer[9] and of Makosza and Wawrzyniewicz,[10] have also been tried, but did not lead to improved yields of adduct.

The rapid conversion of tricyclo[4.4.1.0^{1,6}]undeca-3,8-diene to 1,6-methano[10]annulene by the high potential quinone, DDQ, is yet another illustration of the usefulness of this agent as a means of dehydrogenation of hydroaromatic compounds.[11] If DDQ is not available, it is recommended to aromatize tricyclo[4.4.1.0^{1,6}]undeca-3,8-diene by a bromination–dehydrobromination sequence similar to that described in the synthesis of 1,6-oxido[10]annulene;[12] both aromatization methods give essentially the same yield of 1,6-methano[10]annulene.

The synthesis of 1,6-methano[10]annulene outlined above is an improved version of the method first suggested by Vogel and Roth.[13] 1,6-Methano[10]annulene represents a Hückel-type aromatic $(4n + 2)\pi$-system and is reminiscent of benzene or napthhalene in both its physical and chemical properties.[14] The aromatic nature of the hydrocarbon is borne out most impressively by its proton magnetic resonance spectrum which exhibits an AA'BB'-system for the vinylic protons at relatively low field (2.5–3.2 τ) and a singlet for the bridge protons at relatively high field (10.5 τ). 1,6-Methano[10]annulene may serve as a starting material for the preparation of other molecules of current interest, such as the bicyclo[5.4.1.]dodecapentaenylium ion[14] and benzocyclopropene.[15]

1. Institut für Organische Chemie der Universität Köln, West Germany.
2. W. Hückel and H. Schlee, *Chem. Ber.*, **88**, 346 (1955).
3. W. v. E. Doering and A. K. Hoffmann, *J. Amer. Chem. Soc.*, **76**, 6162 (1954).
4. D. Walker and T. D. Waugh, *J. Org. Chem.*, **30**, 3240 (1965).
5. A. J. Birch and G. Subba Rao, "Advances in Organic Chemistry," Vol. 8, Wiley-Interscience, New York, 1972, p. 1.
6. C. A. Grob and P. W. Schiess, *Helv. Chim. Acta*, **43**, 1546 (1960).
7. P. H. Nelson and K. G. Untch, *Tetrahedron Lett.*, 4475 (1969).
8. E. E. Schweizer and W. E. Parham, *J. Amer. Chem. Soc.*, **82**, 4085 (1960).
9. W. E. Parham and E. E. Schweizer, *J. Org. Chem.*, **24**, 1733 (1959).
10. M. Makosza and M. Wawrzyniewicz, *Tetrahedron Lett.*, 4659 (1969).
11. D. Walker and J. D. Hiebert, *Chem. Rev.*, **67**, 153 (1967).
12. E. Vogel, W. Klug, and A. Breuer, *Org. Syn.*, submitted.
13. E. Vogel and H. D. Roth, *Angew. Chem.*, **76**, 145 (1964).
14. E. Vogel, in "Aromaticity," *Chem. Soc. Spec. Publ.*, W. D. Ollis, ed., No. **21**, p. 113 (1967).
15. E. Vogel, W. Grimme, and S. Korte, *Tetrahedron Lett.*, 3625 (1965).

γ-HYDROXY-α,β-UNSATURATED ALDEHYDES VIA 1,3-BIS(METHYLTHIO)ALLYLLITHIUM: TRANS-4-HYDROXY-2-HEXENAL

(2-Hexenal, 4-hydroxy-)

$$\underset{CH_2-CH-CH_2Cl}{\overset{O}{\triangle}} + 2\,CH_3SH + 2\,NaOH \xrightarrow[\text{below }50°]{\text{methanol}} \underset{CH_2-CH-CH_2}{\overset{SCH_3\;\;OH\;\;SCH_3}{|\quad\;\;|\quad\;\;|}}$$

$$\underset{CH_2-CH-CH_2}{\overset{SCH_3\;\;OH\;\;SCH_3}{|\quad\;\;|\quad\;\;|}} + NaH + CH_3I \xrightarrow[20-25°]{\text{tetrahydrofuran}} \underset{CH_2-CH-CH_2}{\overset{SCH_3\;\;OCH_3\;\;SCH_3}{|\quad\;\;\;|\quad\;\;\;|}}$$

$$\underset{CH_2-CH-CH_2}{\overset{SCH_3\;\;OCH_3\;\;SCH_3}{|\quad\;\;\;|\quad\;\;\;|}} + 2[(CH_3)_2CH]_2NLi \xrightarrow[0°]{\text{tetrahydrofuran-}n\text{-hexane}}$$

$$Li^{\oplus}\left[\underset{CH\;\ominus\;CH}{\overset{SCH_3\;\;CH\;\;SCH_3}{\diagdown\;\;\diagup\diagdown\;\;\diagup}}\right] \xrightarrow[-75-25°]{CH_3CH_2CHO} CH_3CH_2\underset{OH}{\overset{}{\underset{|}{CH}}}-\overset{SCH_3}{\underset{|}{CH}}-CH=\overset{SCH_3}{\underset{|}{CH}}$$

$$CH_3CH_2\underset{OH}{\overset{}{\underset{|}{CH}}}-\overset{SCH_3}{\underset{|}{CH}}-CH=\overset{SCH_3}{\underset{|}{CH}} + HgCl_2 + CaCO_3 \xrightarrow[55°]{\text{tetrahydrofuran-water}}$$

$$CH_3CH_2\underset{OH}{\overset{}{\underset{|}{CH}}}-\overset{H}{\underset{}{C}}=\overset{}{\underset{H}{C}}-CHO$$

Submitted by Bruce W. Erickson[1]
Checked by Susumu Kamata and Wataru Nagata

1. Procedure

Caution! This preparation requires the use of a good hood.

A. *1,3-Bis(methylthio)-2-propanol.* A solution of 44 g. (1.10 moles) of sodium hydroxide in 300 ml. of methanol is placed in a 1-l., four-necked flask equipped with a dry ice reflux condenser, a mechanical stirrer, a thermometer, a gas inlet, and an ice bath. While the solution is stirred and cooled, 50 g. (56 ml., 1.04 moles) of methanethiol (Note 1) is distilled from a lecture bottle into the solution at such a rate that the temperature is maintained below 20°. The gas inlet is then replaced by a 60-ml., glass-stoppered, pressure-equalizing dropping funnel, and, while the reaction mixture is stirred and cooled, 44.4 g. (37.6 ml., 0.48 mole) of epichlorohydrin (Note 2) is added dropwise at such a rate that the temperature is maintained below 50° (Note 3).

The reaction mixture is stirred at 25° for 1 hour, diluted with 500 ml. of water, and extracted with two 200-ml. portions of ether. The combined extract is washed with five 100-ml. portions of water and 100 ml. of saturated aqueous sodium chloride, dried over anhydrous magnesium sulfate (Note 4), filtered, and evaporated at 50° under a pressure of 30 mm. The residual liquid is distilled under reduced pressure without a column (Note 5), and the fraction boiling at 110–111° (7 mm.) or 101–102° (4 mm.) affords 61.0–63.6 g. (84–87%) of 1,3-bis(methylthio)-2-propanol (Notes 6, 7).

B. *1,3-Bis(methylthio)-2-methoxypropane.* In a dry, 1-l., three-necked flask equipped with a mechanical stirrer, a 60-ml. pressure-equalizing dropping funnel, and a thermometer is placed 13.8 g. (0.34 mole) of sodium hydride dispersed in mineral oil (Note 8). The mineral oil is removed by washing the dispersion with five 100-ml. portions of hexane (Note 9). The hexane is removed with a pipet after the sodium hydride has been allowed to settle (Note 10).

When most of the hexane has been removed, 500 ml. of dry tetrahydrofuran (Notes 11 and 12) is added. While the reaction mixture is stirred, 33.4 g. (0.22 mole) of 1,3-bis(methylthio)-2-propanol is added dropwise over 15 minutes. After the evolution of hydrogen ceases, the stirred mixture is cooled with a water bath. When the mixture reaches 20°, 15.0 ml. (0.24 mole) of methyl iodide (Note 2) is added dropwise over 5 minutes. The dropping funnel is replaced by a glass stopper, and the reaction mixture is stirred at 20–25° for 24 hours.

The mixture is concentrated to about 200 ml. at 50° under a pressure of 30 mm., diluted with 200 ml. of ether, and washed with 100 ml. of saturated aqueous sodium chloride, two 100-ml. portions of $0.5M$ aqueous ammonium chloride, and 100 ml. of saturated aqueous sodium chloride. Each of the aqueous washes is extracted with the same 100-ml. portion of ether. The combined ethereal solution is dried over anhydrous magnesium sulfate (Note 4), filtered, and evaporated at 25° under a pressure of 10 mm. The residual material is distilled under reduced pressure without a column (Note 5), and the fraction boiling at 96–97° (9 mm.) or 89–90° (7 mm.) affords 31.4–35.7 g. (86–97%) of 1,3-bis(methylthio)-2-methoxypropane (Notes 13, 14).

C. *1,3-Bis(methylthio)-1-hexen-4-ol.* In a dry 500-ml., three-necked flask containing a Teflon®-coated magnetic stirring bar and equipped with a 100-ml., pressure-equalizing dropping funnel, a thermometer, and a side arm connected to a nitrogen bubbler system is placed 11.65 g. (0.115 mole) of diisopropylamine (Note 15) and 175 ml. of dry tetrahydrofuran (Note 11). The flask is flushed with nitrogen, and a slight positive pressure is maintained during the following operation by a slow stream of nitrogen. The solution is magnetically stirred and cooled to about $-75°$ with a dry ice–acetone bath. After 15 minutes, 77.0 ml. (0.11 mole) of 1.45M solution of n-butyllithium in hexane (Note 16) is added dropwise using the pressure equalizing dropping funnel. While the resulting solution of lithium diisopropylamide is stirred at $-75°$, 9.15 g. (0.055 mole) of 1,3-bis(methylthio)-2-methoxypropane is added through a dropping funnel. The dry ice–acetone bath is replaced by an ice water bath, and the solution is stirred at 0° for 2.5 hours. The solution slowly becomes deep purple as 1,3-bis(methylthio)allyllithium is generated (Note 17).

The solution is stirred and cooled to $-75°$ with a dry ice–acetone bath, and 2.90 g. (0.05 mole) of propionaldehyde (Note 18) is added through a dropping funnel. The solution is stirred at $-75°$ for 5 minutes. The dry ice–acetone bath is replaced by a water bath, and the solution is stirred at 20° for 30 minutes. The solution is diluted with 250 ml. of ether and washed with two 100-ml. portions of 5M aqueous ammonium chloride, two 100-ml. portions of water, and 100 ml. of saturated aqueous sodium chloride. Each of the aqueous washes is extracted with the same 250-ml. portion of ether. The combined ethereal solution is dried over anhydrous magnesium sulfate (Note 4), filtered, and evaporated at 40° under a pressure of 25 mm. The residual liquid is distilled under reduced pressure through a 10 × 0.7 cm., unpacked, vacuum-jacketed column, and the material boiling at 93–98° (2 mm.) furnishes 7.90–7.99 g. (82–83%) of 1,3-bis(methylthio)-1-hexen-4-ol (Notes 19, 20).

D. *trans-4-Hydroxy-2-hexenal.* In a 500-ml., one-necked flask containing a Teflon®-coated magnetic stirring bar is placed 3.85 g. (0.02 mole) of 1,3-bis(methylthio)-1-hexen-4-ol, 80 ml. of tetrahydrofuran (Note 11), and 6.00 g. (0.06 mole) of powdered calcium carbonate. The mixture is stirred, and a solution of 16.4 g. (0.06 mole) of mercuric chloride in 140 ml. of tetrahydrofuran and 40 ml. of water is added. The mixture is stirred and heated at 50–55° with a water bath for 15 hours. The reaction mixture is filtered (Note 21) with suction through a pad of diatomaceous earth in a sintered-glass funnel (Note 22). The filter cake is washed with a mixture of 200 ml. of pentane and 200 ml. of dichloromethane. The combined filtrate is washed with three 100-ml. portions of saturated aqueous sodium chloride. Each of the aqueous washes is extracted with a mixture of 100 ml. of pentane and 100 ml. of dichloromethane. The combined organic solution is dried over anhydrous magnesium sulfate (Note 4), filtered, and evaporated at 25° under a pressure of 25 mm. When almost all the solvent is removed, the residue is slurried with a mixture of 4 ml. of chloroform and 1 ml. of ether as soon as possible. Purification of the product is effected by chromatography on a 3 × 20 cm. column of 50 g. of silicic acid (Note 23) by elution with a 4:1 chloroform–ether mixture. Each fraction (50 ml.) is monitored by thin-layer chromatography, and the fractions containing *trans*-4-hydroxy-2-hexenal uncontaminated with any by-product are collected. The solvent is evaporated at 25° under a pressure of 25 mm. and 1 mm. to yield 1.37–1.41 g. (60–62%) of product, which is pure by gas chromatography (Note 24) and proton magnetic resonance spectroscopy (Note 25). Distillation without a column yields 1.32 g. (58%) of *trans*-4-hydroxy-2-hexenal, b.p. 48–51° (0.01–0.03 mm.), 2,4-dinitrophenylhydrazone, m.p. 198–199° (Note 26).

2. Notes

1. The submitter used methanethiol (b.p. 8°) obtained from Matheson Gas Products. The weight of the methanethiol distilled from the lecture bottle into the reaction flask was measured by difference. The pungent odor of this mercaptan is minimized by slow distillation of the reagent into the flask and by the use of a well-ventilated hood. The checkers used methanethiol (b.p. 8°, d_4^0 0.8948) obtained from Toyo Chemical Industries Co. (Japan) and changed the above procedure as follows: methanethiol distilled from the lecture bottle was first trapped in an ice-cooled graduated flask equipped with a gas inlet and a dry ice reflux condenser, then 56 ml. of the liquid was redistilled into the reaction mixture.

2. The submitter used reagent-grade material obtained from Eastman Kodak Co., and the checkers used reagent-grade material from Kanto Chemical Co., Inc. (Japan).

3. A large quantity of sodium chloride precipitates during addition of the epichlorohydrin.

4. The checkers used anhydrous sodium sulfate as a drying agent instead of anhydrous magnesium sulfate.

5. The checkers used a 15 × 1 cm., unpacked, vacuum-jacketed column for the distillation.

6. For gas chromatographic analysis of the product, the submitter used a 3 ft. × 0.125 in. stainless steel column of 5% LAC-446 (cross-linked diethylene glycol-adipic ester) on Diatoport S (60–80 mesh) swept with prepurified nitrogen at 30 ml. per minute. The retention time was 1.75 minutes at 140°. The checkers used a 1 m. × 4 mm. glass column of 5% OV-17 on Chromosorb W (60–80 mesh) swept with prepurified nitrogen at 90 ml. per minute. The retention time was 2.65 minutes at 150°.

7. The proton magnetic resonance spectrum of the product $[(CH_3^a SCH_2^b)_2 CH^c OH^d]$ in carbon tetrachloride shows peaks at 2.12 (singlet, $6H$, H^a), 2.63 (doublet, $4H$, $J_{b,c} = 6$ Hz., H^b), 3.86 (pentet, $1H$, H^c), and 3.78 p.p.m. (singlet, $1H$, H^d) downfield from internal tetramethylsilane.

8. The sodium hydride was obtained as a 59% dispersion in mineral oil from Metal Hydrides, Inc.

9. The submitter used a reagent-grade mixture of isomeric hexanes (b.p. 68–70°) obtained from Fisher Scientific Co., and the checkers used the same grade mixture of isomeric hexanes obtained from Wako Pure Chemical Industries Ltd. (Japan).

10. About 90% of the mineral oil was removed by this procedure. Because some sodium hydride is lost in the pipet, an excess is initially employed.

11. The submitter used tetrahydrofuran obtained from Fisher Chemical Co. and distilled from lithium aluminium hydride under nitrogen shortly before use. The checkers used tetrahydrofuran obtained from Wako Pure Chemical Industries Ltd. (Japan) and distilled from sodium hydride under nitrogen shortly before use.

12. When dry ether was used in place of dry tetrahydrofuran, only about 15% of the alcohol was methylated in 36 hours.

13. Using the columns described in Note 6, the retention times were 0.67 minute at 140° and 1.00 minute at 120° (submitter) and 2.30 minutes at 150° and 7.05 minutes at 120° (checkers), respectively.

14. The proton magnetic resonance spectrum of the product [$(CH_3^aSCH_2^b)_2CH^cOCH_3^d$] in carbon tetrachloride exhibits absorption at 2.13 (singlet, 6H, H^a), 2.68 (doublet, 4H, $J_{b,c} = 5.4$ Hz., H^b), 3.41 (singlet, 3H, H^d), and 3.50 p.p.m. (pentet, 1H, H^c) downfield from internal tetramethylsilane.

15. The submitter used diisopropylamine from Aldrich Chemical Co., and the checkers used diisopropylamine from Wako Pure Chemical Industries Ltd. (Japan). It was distilled from calcium hydride before use.

16. The submitter used a 1.24M solution of n-butyllithium in pentane obtained from Foot Mineral Co., and the checkers used a 1.45M solution of n-butyllithium in hexane obtained from Wako Pure Chemical Industries Ltd. (Japan). The nominal titer of active alkyl on the bottle agreed well with the value found by titration[2] with 0.80M (submitter) or 0.50M (checkers) solution of 2-butanol in xylene using 1,10-phenanthroline as indicator.

17. Nearly quantitative generation of 1,3-bis(methylthio)allyllithium was proved, as this solution yielded 1,3-bis(methylthio)propene (88–89%) and 1,3-bis(methylthio)-1-butene (89%) by reaction with methanol and methyl iodide, respectively. The checkers found that lithium diisopropylamide can be replaced by n-butyllithium without any trouble for the generation of 1,3-bis(methylthio)allyllithium, simplifying the procedure considerably at least in this particular case. Subsequent reaction with propionaldehyde gave 1,3-bis(methylthio)-1-hexen-4-ol in 85% yield, and no appreciable amount of by-product, such as the addition product of n-butyllithium with propionaldehyde or with the intermediate 1.3-bis(methylthio)propene, was formed.

18. The submitter used propionaldehyde from Aldrich Chemical Co., and the checkers used material obtained from Tokyo Chemical Industry Co. Ltd. (Japan). It was distilled from calcium hydride shortly before use.

19. For gas chromatographic analysis of the products, the submitter used a 3 ft. × 0.125 in. stainless steel column of 5% LAC-446 (crosslinked diethylene glycol-adipic ester) on Diatoport S (60–80 mesh), which was heated at 140° and swept with prepurified nitrogen at 30 ml. per minute. The four isomers were observed as three peaks at retention times of 3.08, 3.69, and 4.07 minutes. The checkers used a 1 m. × 4 mm. glass column of 10% DEGS on Gas Chrome Q (60–80 mesh), which was heated at 200° and swept with prepurified nitrogen at 75 ml. per minute.

The four isomers were observed at retention times of 3.30, 3.85, 4.10, and 5.20 minutes.

20. The proton magnetic resonance spectrum of the product [CH^a(SCH$_3^b$)=CH^{c}CH^d(SCH$_3^e$)—CH^f(OH)CH$_2^g$CH$_3^h$] in carbon tetrachloride exhibits absorption at 0.94 (triplet, 3H, $J_{g,h}$ = 7 Hz., H^h), 1.4 (multiplet, 2H, H^g), 2.00, 2.03, and 2.05 (singlets, 3H, H^e), 2.22 and 2.26 (singlet, 3H, H^b), 2.6–3.8 (multiplet, 2H, H^d and H^f), 5.0–5.8 (multiplet, 1H, H^c), and 5.9–6.3 p.p.m. (multiplet, 1H, H^a) downfield from internal tetramethylsilane. The olefinic multiplets indicate[3] that two *trans* diastereomers and two *cis* diastereomers are present in the ratios 25:23:41:11, respectively.

21. At this stage, the submitter added 2 g. of sodium bicarbonate to buffer the liquid phase near neutrality before filtration. The checkers found that this operation can be omitted without any trouble.

22. Diatomaceous earth is used to avoid clogging the sintered glass filter with the insoluble material in the reaction mixture.

23. To remove the remaining mercuric chloride and some by-products the submitter filtered a carbon tetrachloride solution of the product through a 1-cm. column of Merck silicic acid in an experiment of one-twentieth scale. For the larger-scale preparation the checkers found it necessary to carry out the chromatographic purification using silicic acid (M. Woelm Eshwege, Grade II) as described.

24. Using the same columns described in Note 19, the product shows a retention time of 1.23 minutes at 140° (submitter) and 1.45 minutes at 200° (checkers).

25. The proton magnetic resonance spectrum of the γ-hydroxy-α,β-unsaturated aldehyde [O=CH^{a}CH^b=CH^{c}CH^d(OH)CH$_2^e$CH$_3^f$] in carbon tetrachloride shows absorption at 0.96 (triplet, 3H, H^f), 1.2–1.9 (multiplet, 2H, H^e), 2.22 (singlet, 1H, OH), 4.30 (multiplet, 1H, H^d), 6.23 (doublet of doublets, 1H, H^b), 6.85 (doublet of doublets, 1H, H^c), and 9.49 p.p.m. (doublet, 1H, H^a) downfield from internal tetramethylsilane, with coupling constant J in Hz.: $J_{a,b}$ = 7.6, $J_{b,c}$ = 15.5, $J_{b,d}$ = 1.0, $J_{c,d}$ = 4.3, and $J_{e,f}$ = 7.0. Infrared (CCl$_4$) cm^{-1}: 3615 medium, 3470 medium and broad (OH), 2805 weak, 2720 weak (CH of aldehyde), 1693 strong (C=O), 1640 weak (C=C).

26. *trans*-4-Hydroxy-2-hexenal is unstable and decomposed on standing at room temperature to give some polymeric and dehydrated compounds. The checkers prepared the 2,4-dinitrophenylhydrazone, m.p. 198–199°,[4] for its identification in the usual way.

3. Discussion

1,3-Bis(methylthio)-2-methoxypropane is important as the precursor of 1,3-bis(methylthio)allyllithium,[5] a symmetrical nucleophile that is synthetically equivalent to the presently unknown (and probably intrinsically unstable) 3-lithio derivative of acrolein (Li—CH=CH—CH=O).

Reaction of 1,3-bis(methylthio)-2-methoxypropane with 2 moles of lithium diisopropylamide[5] (or n-butyllithium) effects (a) the elimination of methanol to form 1,3-bis(methylthio)propene and (b) the lithiation of this propene to generate 1,3-bis(methylthio)allyllithium in solution. Its conjugate acid, 1,3-bis(methylthio)propene, can be regenerated by protonation with methanol, and has also been prepared (a) in 31% yield by reaction of methylthioacetaldehyde with the lithio derivative of diethyl methylthiomethylphosphonate,[5] (b) in low yield by acid-catalyzed pyrolysis of 1,1-bis(methylthio)-3-methoxypropane,[6] and (c) in low yield by acid-catalyzed coupling of vinyl chloride with chloromethyl methyl sulfide.[7]

A variety of α,β-unsaturated aldehydes are available by alkylation of 1,3-bis(methylthio)allyllithium and hydrolysis of the product. For

TABLE I
CONVERSION OF EPOXIDES INTO δ-ACETOXY-*trans*-α,β-UNSATURATED ALDEHYDES

Epoxide	Coupling	Acetylation	Hydrolysis	Overall
		Yield, %		
Propylene oxide	97	100	81	78
Cyclopentene oxide	99	92	59	54
Cyclohexene oxide	96	100	85	82

example, *trans*-2-octenal is obtained in 75% yield overall on alkylation with 1-bromopentane and hydrolysis with mercuric chloride.[5]

Addition of 1,3-bis(methylthio)allyllithium to aldehydes, ketones, and epoxides followed by mercuric ion-promoted hydrolysis furnishes hydroxyalkyl derivatives of acrolein[5] that are otherwise available in lower yield by multistep procedures. For example, addition of 1,3-bis-(methylthio)allyllithium to acetone proceeds in 97% yield to give a tertiary alcohol that is hydrolyzed with mercuric chloride and calcium carbonate to *trans*-4-hydroxy-4-methyl-2-pentenal in 41% yield.[5] Addition to an epoxide and hydrolysis affords a δ-hydroxy-α,β-unsaturated aldehyde.[8] Similarly, addition of 1,3-bis(methylthio)allyllithium to an epoxide, acetylation of the hydroxyl group, and hydrolysis with mercuric chloride and calcium carbonate provides a δ-acetoxy-*trans*-α,β-unsaturated aldehyde,[5] as indicated in Table I. Cyclic *cis*-epoxides give aldehydes in which the acetoxy group is *trans* to the 3-oxopropenyl group.

1. The Rockefeller University, New York, New York 10021.
2. S. C. Watson and J. F. Eastham, *J. Organometal. Chem.*, **9**, 165 (1967).
3. B. W. Erickson, Ph. D. Thesis, Harvard University, Cambridge, Mass., 1970.
4. H. Esterbauer and W. Weger, *Monatsh. Chem.*, **98**, 1884 and 1994 (1967).
5. E. J. Corey, B. W. Erickson, and R. Noyori, *J. Amer. Chem. Soc.*, **93**, 1724 (1971).
6. J. Hine, L. G. Mahone, and C. L. Liotta, *J. Org. Chem.*, **32**, 2600 (1967).
7. T. Ichikawa, H. Owatari, and T. Kato, *J. Org. Chem.*, **35**, 344 (1970).
8. E. J. Corey and R. Noyori, *Tetrahedron Lett.*, 311 (1970).

CYCLOPROPYLDIPHENYLSULFONIUM FLUOROBORATE

[Sulfonium, cyclopropyldiphenyl tetrafluoroborate(1-)]

A. $\text{ICH}_2\text{CH}_2\text{CH}_2\text{Cl} + \text{Ph}_2\text{S} + \text{AgBF}_4 \xrightarrow[\text{nitromethane, r.t.} \sim 40°]{}$

$\text{Ph}_2\overset{\oplus}{\text{S}}\text{CH}_2\text{CH}_2\text{CH}_2\text{Cl} \quad \text{BF}_4^{\ominus} + \text{AgI}$

B. $\text{Ph}_2\overset{\oplus}{\text{S}}\text{CH}_2\text{CH}_2\text{CH}_2\text{Cl} \quad \text{BF}_4^{\ominus} + \text{NaH} \xrightarrow[\text{tetrahydrofuran, r.t.}]{}$

$\triangleright\!\!-\!\!\overset{\oplus}{\text{S}}\text{Ph}_2 \quad \text{BF}_4^{\ominus} + \text{NaCl} + \text{H}_2$

Submitted by MITCHELL J. BOGDANOWICZ and BARRY M. TROST[1]
Checked by TSUTOMU AOKI and WATARU NAGATA

1. Procedure

A. *Preparation of 3-Chloropropyldiphenylsulfonium Fluoroborate.* A solution of 93.0 g. (0.50 mole) of diphenyl sulfide (Notes 1 and 2) and 347 g. (1.70 mole) of 1-chloro-3-iodopropane (Notes 2, 3, and 4) in 200 ml. of nitromethane (Note 5) in a 1-l., one-necked flask equipped with a magnetic stirring bar and a nitrogen inlet tube is stirred at room temperature under nitrogen. The flask is shielded from light (Note 6), and 78 g. (0.40 mole) of silver fluoroborate (Note 7) is added in one portion. Initially the temperature rises to 40°, then gradually falls to room temperature. No external cooling is necessary. After 16 hours 200 ml. of methylene chloride is added, and the mixture is filtered through a sintered glass funnel fitted with a pad of 35 g. of Florisil (Note 8). The solid is washed with 100 ml. of methylene chloride, and the methylene chloride portions are combined. This methylene chloride solution is evaporated at reduced pressure until a solid separates and then 1 l. of ether is added to precipitate the sulfonium salt (Note 9). The off-white crystals are collected (Note 10), washed with ether, and dried under reduced pressure at 25°. The yield is 122–140 g. (87–99%), m.p. 103–105° (Note 11).

B. *Preparation of Cyclopropyldiphenylsulfonium Fluoroborate.* A suspension of 118.7 g. (0.339 mole) of 3-chloropropyldiphenylsulfonium fluoroborate (Note 2) in 500 ml. of dry tetrahydrofuran (Note 12) is placed in a 2-l., one-necked flask equipped with a magnetic stirring bar and nitrogen inlet tube under nitrogen. Then 5-g. portions of 55% sodium hydride–mineral oil dispersion (15.2 g., 0.350 mole) are added in 30-minute intervals. The resulting mixture is stirred (Note 13) at room temperature for 24 hours. An aqueous solution of 25 ml. of 48% fluoroboric acid (Note 14), 15 g. of sodium fluoroborate (Notes 7, 15), and 400 ml. of water is added to the well-stirred reaction to destroy residual hydride and swamp out chloride ion (Note 16). After 5 minutes 300 ml. of methylene chloride is added, and the top methylene chloride layer is removed from the lower aqueous layer (Note 17). The methylene chloride solution is then extracted with 100 ml. of water. This water layer is combined with the first aqueous layer, and the combined water layers are extracted with an additional 100 ml. of methylene chloride. The methylene chloride portions are combined, dried over anhydrous sodium sulfate, and evaporated at reduced pressure until precipitation occurs. Addition of 1 l. of ether completes the precipitation of the salt.

The crystals are collected, washed with ether, and recrystallized from hot absolute ethanol (approximately 400 ml.) (Note 18). After drying under reduced pressure the yield is 79.5–88.0 g. (75–83%), m.p. 137–139° (Note 19).

2. Notes

1. Available from Matheson, Coleman and Bell and utilized without further purification. The checkers used reagent grade diphenyl sulfide obtained from Tokyo Kasei Kogyo Co. Ltd., Japan.

2. The checkers carried out the experiment on a half scale and obtained the same results as described by the submitters.

3. Available from K & K Laboratories or may be prepared in 89% yield by the following procedure. To a solution of 393 g. (2.63 moles) of sodium iodide in 1 l. of reagent grade acetone is added 394 g. (2.50 moles) of 1-bromo-3-chloropropane (Aldrich Chemical Co.). After stirring 2 hours at room temperature, the mixture is filtered, the sodium bromide is washed with acetone, and the acetone is evaporated at reduced pressure. A dark iodine color is present along with some solid sodium salts. The oil is dissolved in ether, and the solution is washed with a 10% aqueous sodium thiosulfate solution. The ethereal layer is separated, dried over anhydrous sodium sulfate, and evaporated at reduced pressure to yield 454 g. of an oil that can be used without further purification.

4. An excess of 1-chloro-3-iodopropane must be employed to compete effectively with the diphenyl sulfide for complexation with silver fluoroborate.

5. Available from Aldrich Chemical Co. and used without further purification. Methylene chloride may be substituted for the nitromethane. The checkers used reagent grade nitromethane available from Tokyo Kasei Kogyo Co. Ltd., Japan.

6. The flask is wrapped with aluminium foil to prevent decomposition of the silver salts.

7. Available from Ozark Mahoning Corp.

8. A pad of Florisil is employed to facilitate removal of the suspended silver salts.

9. An oil separated initially. Vigorous shaking of the mixture to extract the excess starting material out of the oily sulfonium salt layer induces crystallization.

10. The crystals obtained by the checkers were light brown at this stage, but could be purified by the following procedure.

11. The material is normally utilized directly without further purification. If the solid is very gray, it may be recrystallized. For the recrystallization the salt is dissolved in hot 95% ethanol (approximately 350 ml. per 100 g. of salt) containing decolorizing carbon, filtered rapidly, and the clear supernatant liquid is allowed to cool in a freezer ($-20°$). In this way, white crystals, m.p. 106–107°, may be obtained with nearly quantitative recovery. The checkers obtained the purified material of m.p. 108–109° with 95% recovery and used this material for the next step. The purified material has the following spectral data; ultraviolet (95% ethanol) nm. max (ϵ): 236 shoulder (13,200), 262 (2200), 268 (2680), 275 (1010); infrared (Nujol) cm^{-1}: 3090 weak, 3060 weak (aromatic CH), 1580 weak (C=C); proton magnetic resonance (CDCl$_3$), first-order analysis, δ in p.p.m.: 7.5–8.2 (multiplet 10H, H^a), 4.3 (poorly resolved triplet 2H, H^b), 3.75 (triplet 2H, H^d), 2–2.5 (multiplet 2H, H^c); coupling constant J in Hz.: $J_{bc} = 8$, $J_{cd} = 6.5$.

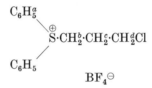

12. The tetrahydrofuran was dried by distilling from lithium aluminium hydride and then from sodium benzophenone ketyl (generated by adding small pieces of sodium metal and benzophenone) directly into the reaction flask. A blue-black color of the ketyl solution indicates dryness. The checkers purified tetrahydrofuran by distillation from sodium hydride dispersion under nitrogen, and used it immediately.

13. The checkers found that efficient stirring is essential for successful results.

14. Since 48% fluoroboric acid was not available in Japan, the checkers used 42% fluoroboric acid obtained from Wako Pure Chemicals Co. Ltd. and obtained the same result as described by the submitters.

15. The checkers prepared sodium fluoroborate by neutralization of an ice-cold, aqueous, 42% solution of fluoroboric acid with an equivalent amount of sodium carbonate and addition of dry ethanol to the reaction mixture to effect complete crystallization of the product. The crystals were purified by washing with ethanol, and the purified material was obtained in 80% yield.

16. Normally no observable effect occurs upon this addition. Gas evolution with a slight exotherm indicates incomplete reaction.

17. The density of the methylene chloride and water layers are nearly equal. Thus, sometimes upon initial mixing, the methylene chloride starts out on the bottom, but the layers reverse on shaking. However, on occasion, the desired methylene chloride layer is found in fact to be the bottom one. It is therefore advisable to check the layers by addition of either water or methylene chloride. The checkers found that the methylene chloride was on the bottom in the two experiments.

18. Ether may be added to the cold ethanol solution before filtration to insure complete precipitation.

19. The purified material has the following spectral data; ultraviolet (95% ethanol) nm. max (ϵ): 235 shoulder (12,200), 261 (1800), 267 (2200), 274 (1700); infrared (Nujol) cm^{-1}: 3100 weak, 3045 weak (aromatic CH), 1582 weak (C=C); proton magnetic resonance (CDCl$_3$), first-order analysis, δ in p.p.m.: 7.5–8.1 (multiplet 10H, H^a), 3.5–3.9 (multiplet 1H, H^b), 1.4–1.75 (multiplet 4H, H^c).

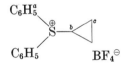

3. Discussion

The utility of sulfur ylides in organic synthesis demands methods for the efficient preparation of the precursor sulfonium salts.[2] Among the salts, the diphenylsulfonium moiety provides the ability to generate the higher alkylides unambiguously. However, the low nucleophilicity of the sulfur of diphenyl sulfide dictated the need for exceptionally reactive alkylating agents. Oxonium salts,[3] dialkoxycarbonium salts,[4] and fluorosulfate esters[5] are capable of achieving such alkylations; however, the unavailability of such alkylating agents except for the very simple alkyl groups (e.g., methyl and ethyl) does not allow generalization. On the other hand, alkyl halides complexed to silver salts form powerful alkylating agents and allow utilization of a wide range of alkyl halides susceptible to S$_N$2 displacement.[2,6] Although alkyl bromides may be employed, alkyl iodies are preferred. The latter are normally available in excellent yields from sulfonate esters, chlorides, or bromides by reaction with sodium iodide. Polyhalides may be employed without complications—reaction occurring preferably at a primary rather than secondary center.

While sulfonium ylides do not normally undergo alkylations (except with reactive alkylating agents such as methyl iodide[7]), they do undergo intramolecular alkylation (cyclization) rather efficiently. The present procedure describes the synthesis of a particularly interesting reagent, cyclopropyldiphenylsulfonium fluoroborate.[8] The ylide derived from this salt effects many different synthetic transformations which include facile syntheses of cyclobutanones,[8] γ-butylrolactones,[9] and specifically substituted cyclopentanones[10] from aldehydes and ketones and spiropentanes from α,β-unsaturated carbonyl partners.[8,11]

1. Department of Chemistry, University of Wisconsin, Madison, Wisconsin 53706
2. E. J. Corey and M. Chaykovsky, *J. Amer. Chem. Soc.*, **87**, 1353 (1965); E. J. Corey and W. Oppolzer, *J. Amer. Chem. Soc.*, **86**, 1899 (1964); V. Franzen and H. E. Driesin, *Chem. Ber.*, **96**, 1881 (1963); V. Franzen, H. J. Schmidt, and C. Mertz, *Chem. Ber.*, **94**, 2942 (1961).
3. H. Meerwein, D. Delfs, and H. Morschel, *Angew. Chem.*, **72**, 927 (1960).
4. H. Meerwein, K. Bodenbenner, P. Borner, F. Kunert, and K. Wunderlich, *Justus Liebigs Ann. Chem.*, **632**, 38 (1960); H. Meerwein, P. Laasch, R. Mersch, and J. Spille, *Chem. Ber.*, **89**, 209 (1965); S. Kabuss, *Angew. Chem. Int. Ed. Engl.* **5**, 675 (1966).
5. M. G. Ahmed, R. W. Alder, G. H. James, M. L. Sinnott, and M. C. Whiting, *Chem. Commun.*, 1533 (1968); M. G. Ahmed and R. W. Alder, *Chem. Commun.*, 1389 (1969).
6. Mercury salts (J. van der Veen, *Rec. Trav. Chim. Pays-Bas*, **84**, 540 (1965)) and antimony salts (G. A. Olah, J. R. DeMember, R. H. Schlosberg, and Y. Halpern, *J. Amer. Chem. Soc.*, **94**, 156 (1972) and references therein) may also be employed.
7. E. J. Corey, M. Jautelat, and W. Oppolzer, *Tetrahedron Lett.*, 2325 (1967).
8. B. M. Trost and M. J. Bogdanowicz, *J. Amer. Chem. Soc.*, **93**, 3773 (1971); *J. Amer. Chem. Soc.*, **95**, 5321 (1973).
9. M. J. Bogdanowicz and B. M. Trost, *Tetrahedron Lett.*, 923 (1973).
10. B. M. Trost and M. J. Bogdanowicz, *J. Amer. Chem. Soc.*, **95**, 289, 5311 (1973).
11. For a related reagent see C. R. Johnson, G. F. Katekar, R. F. Huxol, and E. R. Janiga, *J. Amer. Chem. Soc.*, **93**, 3771 (1971).

TRIMETHYLENE DITHIOTOSYLATE AND ETHYLENE DITHIOTOSYLATE

(1,3-Propanedithiol di-p-toluenesulfonate and 1,2-Ethanedithiol di-p-toluenesulfonate)

$$KOH + H_2S \xrightarrow[0°]{H_2O} KHS + H_2O$$

$$CH_3C_6H_4SO_2Cl + 2\ KHS \xrightarrow[55-60°]{H_2O} CH_3C_6H_4SO_2SK + KCl + H_2S$$

$$2\ CH_3C_6H_4SO_2SK + Br(CH_2)_3Br \xrightarrow[\text{reflux}]{C_2H_5OH,\ KI}$$
$$CH_3C_6H_4SO_2S(CH_2)_3SSO_2C_6H_4CH_3 + 2\ KBr$$

$$2\ CH_3C_6H_4SO_2SK + Br(CH_2)_2Br \xrightarrow[\text{reflux}]{C_2H_5OH,\ KI}$$
$$CH_3C_6H_4SO_2S(CH_2)_2SSO_2C_6H_4CH_3 + 2\ KBr$$

Submitted by R. B. WOODWARD,[1] I. J. PACHTER, and MONTE L. SCHEINBAUM[2]
Checked by P. A. ROSSY and S. MASAMUNE

1. Procedure

A. *Potassium Thiotosylate. Caution!* This procedure should be carried out in a hood to avoid inhalation of hydrogen sulfide.

A solution of 64.9 g. (1 mole) of 86.5% potassium hydroxide (Note 1) in 28 ml. of water is cooled in an ice bath, saturated with hydrogen sulfide, and flushed with nitrogen to ensure complete removal of excess hydrogen sulfide (Notes 2 and 3). The freshly prepared potassium hydrosulfide solution is diluted with 117 ml. of water and stirred under nitrogen at 55–60°. Then 95.3 g. (0.5 mole) of finely ground tosyl chloride (Note 3) is introduced in small portions at a uniform rate so that the reaction temperature is maintained at 55–60° (Note 2). A mildly exothermic reaction ensues, and the solution becomes intensely yellow. After about 90 g. of tosyl chloride has been introduced, the yellow color disappears, and the dissolution of the chloride ceases. The reaction mixture is rapidly filtered with suction through a warmed funnel, and the filtrate is cooled several hours at 0–5°. The crystals of potassium thiotosylate are filtered, dissolved in 200 ml. of hot 80% ethanol, filtered hot to remove traces of sulfur, and cooled several hours at 0–5°. The recrystallized salt is filtered and air-dried to provide 48–55 g. (42–49%) of white crystals.

B. *Trimethylene Dithiotosylate.* To 150 ml. of 95% ethanol containing 10–20 mg. of potassium iodide is added 40 g. (0.177 mole) of potassium thiotosylate and 20 g. (0.10 mole) of trimethylene dibromide (Note 4). The mixture is refluxed with stirring for 8 hours in the dark and under nitrogen. The reaction mixture is cooled to ambient temperature, diluted with an equal volume of cold water, and agitated. After decantation of the supernatant liquid, the residual honey-like layer of product is washed with three 200-ml. portions of cold water, once with 100 ml. of cold 95% ethanol, and once with 100 ml. of cold absolute ethanol. The crude product (Note 5) is dissolved in 10 ml. of acetone, diluted with 80 ml. of hot absolute ethanol, and stirred under nitrogen at 0°. The oil which separates is redissolved by the addition of a minimum amount (*ca.* 5 ml.) of acetone. Seed crystals are introduced (Note 6), and the mixture is stirred for 1 hour at 0° under nitrogen and stored at −30° for several hours. The microcrystalline product is collected by filtration and weighs 20.2 g., m.p. 63.5–65.0°. Three recrystallizations from nine parts (180 ml.) of ethanol give white needles that weigh 17.2 g. (41%) and melt at 66–67°. During the recrystallizations some of the material oils out when the solution is cooled to room temperature. The supernatant liquid is decanted, seeded, and stored at −30° for several hours. The oil is not further purified. The recrystallized material is chemically pure for further use.[3] Its properties can be compared with those of tosyltrimethylene thiotosylate, isolated from contaminated samples of trimethylene thiotosylate (Notes 7 and 8).

C. *Ethylene Dithiotosylate.* To 200 ml. of ethanol containing 10–20 mg. of potassium iodide is added 45.3 g. (0.2 mole) of potassium thiotosylate and 18.8 g. of ethylene dibromide. The mixture is refluxed with stirring for 8 hours in the dark and under a nitrogen atmosphere. The solvent is removed, and the resulting white solid is washed with a mixture of 80 ml. of ethanol and 150 ml. of water. After decantation, the solid is washed three times with 50-ml. portions of water and then recrystallized from approximately 150 ml. of ethanol to yield 28.7 g. of crude product, m.p. 72–75°. Three recrystallizations from a mixture of ethyl acetate and ethanol afford 24 g. (60%) of white crystals, m.p. 75–76° (Note 9).

2. Notes

1. Potassium hydroxide pellets of reagent grade commonly available, such as that from Fisher Scientific Company, contain 10–15% of water. The checkers used the amount calculated on the basis of 86.5%, as specified.

2. Hydrogen sulfide is undesirable because its presence can lead to the formation of potassium p-toluenesulfinate. The latter can be formed by the desulfurization of thiotosylate by hydrogen sulfide generated in the reaction of potassium hydrosulfide with tosyl chloride. Attention should be directed toward control of the reaction temperature so that hydrogen sulfide is rapidly removed, thereby ensuing survival of the S—S bond of the thiotosylate. p-Toluenesulfinate ion can displace bromide to form stable sulfones which are less soluble in common solvents, such as benzene, than trimethylene dithiotosylate. Therefore, purification of the dithiotosylate contaminated with the sulfones is difficult to achieve by means of fractional recrystallization.

3. The p-toluenesulfonyl chloride should be free of p-toluenesulfonic acid, otherwise potassium p-toluenesulfonate will be formed and will result in the formation of tosylates, rather than thiotosylates. The reagent used by the checkers was obtained from British Drug Houses Ltd. and was purified according to the following procedure.[4] A benzene solution of the tosyl chloride was washed with 5% aqueous sodium hydroxide, dried with magnesium sulfate, and then distilled under reduced pressure, b.p. 146° (15 mm.).

4. Trimethylene dibromide available from Eastman Organic Chemicals, was distilled prior to use (b.p. 167–168°).

5. The checkers found that the crude oily product crystallizes after storage for a few days under nitrogen at −30°. Some of this solid was saved to be used as seed crystals.

6. The submitters reported that seed crystals were obtained by column chromatography, using 40 parts by weight of Woelm neutral alumina (activity grade one) and benzene elution. The center cuts of m.p. 65° or higher were combined and recrystallized from nine parts of ethanol to give white needles, m.p. 67°. Two recrystallizations of chromatographed trimethylene dithiotosylate gave material with m.p. 67.5°.

7. The purified trimethylene dithiotosylate exhibits infrared bands (CHCl$_3$ solution) at 3030 (w), 2930 (w), 1590 (w), 1490 (w), 1440 (w),

1410 (w), 1325 (s), 1300 (m), 1180 (w), 1140 (s), 1075 (s), 1015 (w), and 810 (m) cm^{-1}. The proton magnetic resonance spectrum (CDCl$_3$ solution) included signals at δ1.98 (quintet, 2H, CH$_2$CH_2CH$_2$, J = 7 Hz.), 2.43 (singlet, 3H, CH_3), 2.97 (triplet, 4H, CH_2CH$_2$CH_2, J = 7 Hz.), 7.30 (doublet, 4H, J = 9 Hz.), and 7.75 p.p.m. (doublet, 4H, J = 9 Hz.). Analysis calculated for C$_{17}$H$_{20}$O$_4$S$_4$: C, 49.01; H, 4.84; S, 30.79. Found: C, 49.13; H, 4.81; S, 30.51 (submitters). Found: C, 48.71; H, 4.64; S, 30.45 (checkers). The checkers found that the product exhibited a single peak on a 3 ft. × 1/8 in. Waters Associates Analytical Liquid Chromatographic column, packed with Durapak-Carbowax 400/Poracil C. Chloroform was used as the eluting solvent.

8. The submitters succeeded in isolating tosyltrimethylene thiotosylate, m.p. 92° from contaminated samples of trimethylene dithiotosylate (Note 2). The infrared spectrum (CHCl$_3$ solution) is almost identical with that of trimethylene dithiotosylate. The material has proton magnetic resonance peaks (CDCl$_3$ solution) at δ2.06 (quintet, 2H, CH$_2$CH_2CH$_2$, J = 12 Hz.), 2.43 (singlet, 3H, CH_3), 3.08 (triplet, 4H, CH_2CH$_2$CH_2, J = 12 Hz.), 7.29 (overlapping doublets, 4H, J = 12 Hz.), and 7.72 p.p.m. (overlapping doublets, 4H, J = 12 Hz.). Analysis calculated for C$_{17}$H$_{20}$O$_4$S$_3$: C, 53.09; H, 5.24; S, 25.01. Found: C, 53.04; H, 5.23; S, 25.0.

9. The proton magnetic resonance spectrum (CDCl$_3$ solution) has the following absorptions, δ2.47 (singlet, 6H, CH_3), 3.31 (singlet, 4H, CH_2), 7.48 (complex multiplet, 4H, J = 9 Hz.) and 7.97 p.p.m. (doublet, 4H, J = 9 Hz.). Analysis calculated for C$_{16}$H$_{18}$S$_4$O$_4$: C, 47.73; H, 4.51; S, 31.86. Found: C, 47.89; H, 4.44; S, 32.22.

3. Discussion

Although it has been long known that trimethylene dithiotosylate can be prepared by the reaction of thiotosylate ion with trimethylene dibromide,[5] various difficulties are associated with the preparation. These problems are to a considerable extent related to the mode of preparation and the resultant purity of potassium thiotosylate. The thiotosylate salt must be free of tosylate and p-toluenesulfinate impurities, otherwise side products such as tosylates or sulfones will form. One such by-product, tosyltrimethylene thiotosylate, CH$_3$C$_6$H$_4$SO$_2$(CH$_2$)$_3$SSO$_2$C$_6$H$_4$CH$_3$, m.p. 92°, was isolated from contaminated samples of trimethylene dithiotosylate. It is products such as these, that make crystallization of the dithiotosylate difficult. The procedure described herein serves as a reliable technique for minimizing these experimental difficulties.

Trimethylene dithiotosylate can react with activated methylene groups, enamines, or hydroxyethylene derivatives of carbonyl compounds to form dithiane derivatives. Ethylene dithiotosylate undergoes similar reactions to form dithiolanes.[3,6]

1. Department of Chemistry, Harvard University, Cambridge, Massachusetts 02138.
2. Department of Chemistry, East Tennessee State University, Johnson City, Tennessee 37601
3. R. B. Woodward, I. J. Pachter, and M. L. Scheinbaum, *Org. Syn.*, **54**, 37 (1974).
4. D. D. Perrin, W. L. F. Armarego, and D. R. Perrin, "Purification of Laboratory Chemicals," Pergamon Press, London, 1966, p. 268.
5. (a) J. C. A. Chivers and S. Smiles, *J. Chem. Soc.*, 697 (1928): (b) L. G. S. Brooker and S. Smiles, *J. Chem. Soc.*, 1723 (1926).
6. R. B. Woodward, I, J. Pachter, and M. L. Scheinbaum, *J. Org. Chem.* **36**, 1137 (1971).

2,2-(ETHYLENEDITHIO)CYCLOHEXANONE

(1,4-Dithiaspiro[4,5]decan-6-one)

$$\text{cyclohexanone-CHOH} + CH_3C_6H_4SO_2S(CH_2)_2SSO_2C_6H_4CH_3 \xrightarrow[CH_3OH\ 65°]{KOCOCH_3} \text{product}$$

Submitted by R. B. WOODWARD[1], I. J. PACHTER, and M. L. SCHEINBAUM[2]
Checked by J. G. GREEN and S. MASAMUNE

1. Procedure

A 300-ml., one-necked flask is equipped with a reflux condenser, to the top of which a nitrogen inlet tube is attached. The flask is charged with 3.85 g. (0.03 mole) of 2-hydroxymethylenecyclohexanone (Note 1), 10 g. (0.025 mole) of ethylene dithiotosylate (Note 2), and 10 g. of potassium acetate in 150 ml. of methanol, and the mixture is refluxed for 3 hours with stirring and under nitrogen. The solvent is removed from the reaction mixture on a rotary evaporator, and the residue is extracted with three 50-ml. portions of ether. The combined ethereal extracts are washed with cold aqueous $2N$ sodium hydroxide (Note 3) until the aqueous layer is basic to litmus and then with 50 ml. of saturated aqueous sodium chloride. The ethereal layer is dried over anhydrous magnesium sulfate; the drying agent is removed by filtration, and the ethereal solution is concentrated on a rotary evaporator. The oily residue is diluted with 1 ml. of benzene and 3 ml. of cyclohexane. This solution is transferred to a chromatographic column (14 × 2 cm.) prepared with 50 g. of alumina (Note 4) and a 3:1 mixture of cyclohexane and benzene. With this solvent system the desired product moves with the solvent front, and the first 100 ml. of eluent contains 85% of the total product. Further elution with approximately 100 ml. of the same solvent mixture removes the rest of the material, and then a second component begins to come off. Evaporation of the solvent from the combined 200 ml. of eluent leaves an oily residue which crystallizes on standing to yield 2.76–3.04 g. (57–64%) of crude 2,2-(ethylenedithio)-cyclohexanone. Recrystallization from approximately 50 ml. of pentane affords 2.1–2.6 g. (45–55%) of needles, m.p. 56–57° (Note 5).

2. Notes

1. 2-Hydroxymethylenecyclohexanone was prepared by both the submitters and checkers by a procedure similar to, but slightly modified from, that described by C. Ainsworth in *Org. Syn.*, Coll. Vol. **4**, 536 (1963). To a cooled (ice bath), stirred suspension of 10.2 g. (0.2 mole) of commercial sodium methoxide in 75 ml. of anhydrous benzene in a nitrogen atmosphere was added dropwise but rapidly (*ca.* 2 minutes) a mixture of 9.8 g. (0.1 mole) of distilled cyclohexanone and 14.8 g. (0.2 mole) of distilled ethyl formate. After addition, the reaction was allowed to warm to room temperature and left overnight. Ice water (100 ml.) was added to the resulting suspension. The aqueous layer was separated, and the benzene layer was washed three times with 50 ml. of cold aqueous $0.1N$ sodium hydroxide. The aqueous layers were combined, and the product was isolated according to the procedure referenced above. This modified version provided slightly higher yields of the product than that recorded in *Org. Syn.*, and the ease of handling sodium methoxide, compared with sodium metal, is advantageous.

2. Ethylene dithiotosylate of m.p. 73–73.5°[3] is employed.

3. Treatment with alkali removes the various acidic by-products and their salts (acetate, sulfinate, and formate) and also serves to hydrolyze and remove unreacted starting materials.

4. The checkers used "Aluminium Oxide" purchased from J. T. Baker Chemical Company.

5. The proton magnetic resonance spectrum of the product ($CDCl_3$ solution) is: δ 3.30 (singlet, $4H$), 2.73 (multiplet, $2H$), 2.42 (multiplet, $2H$), and 1.83 (multiplet, $4H$).

3. Discussion

The procedure for the preparation of a dithiolane from a hydroxymethylene derivative of a ketone and ethylene dithiotosylate (ethane-1,2-dithiol di-*p*-toluenesulfonate) can be varied to produce dithianes when the latter reagent is replaced by trimethylene dithiotosylate.[3,4] The dithiotosylates also react with enamine derivatives to produce dithiaspiro compounds.[4,5]

1. Department of Chemistry, Harvard University, Cambridge, Massachusetts 02138.
2. Department of Chemistry, East Tennessee State University, Johnson City, Tennessee 37601.
3. R. B. Woodward, I. J. Pachter, and M. L. Scheinbaum, *Org. Syn.*, **54**, 33 (1974).
4. R. B. Woodward, I. J. Pachter, and M. L. Scheinbaum, *J. Org. Chem.*, **36**, 1137 (1971).
5. R. B. Woodward, I. J. Pachter, and M. L. Scheinbaum, *Org. Syn.*, **54**, 39 (1974).

2,2-(TRIMETHYLENEDITHIO)CYCLOHEXANONE

(1,5-Dithiaspiro[5,5]undecan-7-one)

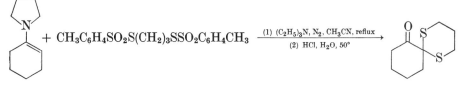

Submitted by R. B. Woodward,[1] I. J. Pachter, and M. L. Scheinbaum[2]
Checked by G. S. Bates and S. Masamune

1. Procedure

A. *1-Pyrrolidinocyclohexene.*[3] A solution of 29.4 g. (0.3 mole) of cyclohexanone and 28.4 g. (0.4 mole) of pyrrolidine in 150 ml. of benzene is placed in a 500-ml., one-necked flask to which a Dean–Stark water separator is attached. The solution is refluxed under a nitrogen atmosphere until the separation of water ceases (Note 1). The excess pyrrolidine and benzene are removed from the reaction mixture on a rotary evaporator. The resulting residue is stored under refrigeration and distilled just before use in the next step to provide 44.6 g. of 1-pyrrolidinocyclohexene, b.p. 76–77° (0.5 mm.), 105–106° (13 mm.).

B. *2,2-(Trimethylenedithio)cyclohexanone.* A solution of 3.02 g. (0.02 mole) of freshly distilled 1-pyrrolidinocyclohexene, 8.32 g. (0.02 mole) of trimethylene dithiotosylate[4] (Note 2), and 5 ml. of triethylamine (Note 3) in 40 ml. of anhydrous acetonitrile (Note 4), is refluxed for 12 hours in a 100-ml., round-bottom flask under a nitrogen atmosphere. The solvent is removed under reduced pressure on a rotary evaporator, and the residue is treated with 100 ml. of aqueous $0.1 N$ hydrochloric acid for 30 minutes at 50° (Note 5). The mixture is cooled to ambient temperature and extracted with three 50-ml. portions of ether. The combined ether extracts are washed with aqueous 10% potassium bicarbonate solution (Note 6) until the aqueous layer remains basic to litmus, and then with saturated sodium chloride solution. The ethereal solution is dried over anhydrous sodium sulfate, filtered, and concentrated on a rotary evaporator. The resulting oily residue is diluted with 1 ml. of benzene and then with 3 ml. of cyclohexane. The solution is poured into a chromatographic column (13 × 2.5 cm.), prepared with 50 g. of alumina (Note 7) and a 3:1 mixture of cyclohexane and benzene. With this solvent system, the desired product moves with the solvent front, and the first 250 ml. of eluent contains 95% of the total product. Elution with a further 175 ml. of solvent removes the remainder. The combined fractions are evaporated, and the pale yellow, oily residue crystallizes readily on standing. Recrystallization of this material from pentane gives 1.82 g. of white crystalline 2,2-(trimethylenedithio)cyclohexanone, m.p. 52–55° (45% yield) (Note 8).

2. Notes

1. The time required for this operation generally is 3.5–5 hours.
2. Trimethylene dithiotosylate[4] of m.p. 66–67° is employed.

3. Eastman white label triethylamine is distilled from sodium hydroxide.

4. Fisher Reagent acetonitrile is distilled from phosphorus pentoxide.

5. Treatment with the dilute acid effects aqueous extraction of pyrrolidine and hydrolysis of unreacted dithiotosylate and enamine starting materials.

6. Bicarbonate washing ensures removal of the sulfonic and sulfinic acids.

7. The checkers used "Aluminium Oxide" purchased from J. T. Baker Chemical Company.

8. The proton magnetic resonance spectrum of the product ($CDCl_3$ solution) exhibits multiplets in the region δ 1.65–2.45. The infrared spectrum ($CHCl_3$ solution) shows peaks at 2980 (m), 2940 (s), 2870 (m), 1690 (s), 1445 (m), 1420 (m), 1120 (m), 1110 (m), and 910 (s) cm^{-1}.

3. Discussion

The preparation of dithianes from enamines by reaction with trimethylene dithiotosylate (propane-1,3-dithiol di-*p*-toluenesulfonate) has been applied with enamines derived from cholestan-3-one, acetoacetic ester, and phenylacetone.[5] Reactions of trimethylene dithiotosylate with hydroxymethylene derivatives of ketones also give rise to dithianes; thus the hydroxymethylene derivative of cholest-4-en-3-one can be converted to 2,2-(trimethylenedithio)cholest-4-en-3-one.[6] 1,3-Dithiolanes are obtained in a similar manner by reaction of ethylene dithiotosylate[4] with the appropriately activated substrate.[5,7]

1. Department of Chemistry, Harvard University, Cambridge, Massachusetts 02138.
2. Department of Chemistry, East Tennessee State University, Johnson City, Tennessee 37601.
3. L. A. Cohen and B. Witkop, *J. Amer. Chem. Soc.*, **77**, 6595 (1955); G. Stork, A. Brizzolara, H. Landesman, J. Szmuszkovicz, and R. Terrell, *J. Amer. Chem. Soc.*, **85**, 207 (1963).
4. R. B. Woodward, I. J. Pachter, and M. L. Scheinbaum, *Org. Syn.*, **54**, 33 (1974).
5. R. B. Woodward, I. J.Pachter, and M. L. Scheinbaum, *J. Org. Chem.*, **36**, 1137 (1971).
6. R. B. Woodward, A. A. Patchett, D. H. R. Barton, D. A. J. Ives, and R. B. Kelley, *J. Chem. Soc.*, 1131 (1957).
7. R. B. Woodward, I. J. Pachter, and M. L. Scheinbaum, *Org. Syn.*, **54**, 37 (1974).

ALDEHYDES FROM 4,4-DIMETHYL-2-OXAZOLINE AND GRIGNARD REAGENTS: o-ANISALDEHYDE

(o-Methoxybenzaldehyde)

Submitted by R. S. Brinkmeyer, E. W. Collington, and A. I. Meyers[1]
Checked by R. E. Ireland and R. R. Schmidt, III

1. Procedure

In a 1-l., three-necked, round-bottomed flask equipped with a 500-ml. dropping funnel (Note 1), a mechanical stirrer, and an argon inlet tube is placed 80 g. (0.33 mole) of N,4,4-trimethyl-2-oxazolinium iodide (Note 2). The system is flushed with argon; 150 ml. of dry tetrahydrofuran (Note 3) is added, and the stirred suspension is cooled in an ice bath. Meanwhile, to a cooled solution of freshly prepared o-methoxyphenylmagnesium bromide (0.414 mole) (Note 4) is added 146 ml. (0.828 mole) of dry hexamethylphosphoramide (Note 5). This solution is then transferred under an argon atmosphere to the 500-ml. dropping funnel with the aid of an argon-pressurized siphon. The solution is slowly run into the cooled suspension, whereupon the methiodide salt dissolves. When the addition is complete, the reaction mixture is stirred at room temperature overnight.

The suspension is slowly poured into 600 ml. of saturated ammonium chloride solution which has previously been cooled to 10–15°, and the mixture is extracted with three 250-ml. portions of ether. Concentration of the combined extracts gives the crude addition product (Note 6).

The crude product is added to 200 ml. of ice-cold water and quickly acidified with cold 3N hydrochloric acid. The acidic solution is rapidly extracted with 300 ml. of cold hexane, and the extract is discarded. The aqueous solution is then made basic by the addition of 20% aqueous sodium hydroxide solution (Note 7), and the mixture is extracted with three 250-ml. portions of ether. Concentration of the combined ethereal extracts affords 70–75 g. of crude oxazolidine (Note 8).

In a 1-l., round-bottomed flask is placed 72 g. of the crude oxazolidine in 600 ml. of water, and 201.6 g. (1.6 moles) of hydrated oxalic acid is added. The mixture is then heated under reflux for 1 hour, cooled, treated with 600 ml. of water to dissolve precipitated oxalic acid, and extracted with three 100-ml. portions of ether. The combined ethereal extracts are washed with 50 ml. saturated sodium bicarbonate solution and then dried over anhydrous potassium carbonate. Concentration of the ethereal solution gives 30–35 g. of crude aldehyde. Distillation of this material at 70–75° (1.5 mm.) gives pure o-anisaldehyde (22.8–26.3 g.; 51–59%), m.p. 35.5–38° (Note 9).

2. Notes

1. The dropping funnel should be equipped so that the transfer of the Grignard reagent to it can be carried out under a positive nitrogen pressure.

2. 4,4-Dimethyl-2-oxazoline is commercially available from Columbia Organic Chemicals, 912 Drake Street, Columbia, South Carolina, or may be prepared as follows. In a 250-ml., three-necked flask is placed 89.14 g. (1.0 mole) of 2-amino-2-methyl-1-propanol, and the flask is cooled in an ice bath. The amine is carefully neutralized with 52.3 g. (1.0 mole) of 90.6% formic acid over a 1-hour period. A magnetic stirring bar is added, the flask is fitted with a short path distillation head, and the reaction mixture is placed in a silicon oil bath which is rapidly heated to 220–250°. The azeotropic mixture of water and oxazoline distills over a period of 2–4 hours and is collected in an ice-cooled flask containing ether. The aqueous layer is separated, saturated with sodium chloride, and extracted with three 50-ml. portions of ether. The combined ethereal extracts are dried over potassium carbonate, filtered to remove the drying agent, and the ether is removed at 35–40° at atmospheric pressure. The 4,4-dimethyl-2-oxazoline is collected as the temperature rises above 85°. The yield is 56.7–62.7 g. (57–63%) of a colorless mobile liquid, b.p. 99–100° (758 mm. Hg).

The checkers found that if the azeotropic mixture is distilled more slowly from the reaction mixture at a pot temperature of 175–195° the yield is greatly reduced and large amounts of polymeric material are formed.

N,4,4-Trimethyl-2-oxazolinium iodide is prepared by adding 49.5 g. (0.5 mole) of 4,4-dimethyl-2-oxazoline to an excess of cold methyl iodide (182 g., 80 ml., 1.28 moles) in a 500-ml. flask and stirring at room temperature under argon for 20 hours. The light brown solid is filtered with the aid of suction and then dissolved in 350 ml. of dry acetonitrile. The methiodide salt is precipitated by the addition of 750 ml. of dry ether to this acetonitrile solution. The purified salt is again filtered with the aid of suction, and the white solid is washed with 250 ml. of dry ether and finally dried under vacuum. This gives 96 g. (80%) of the methiodide, m.p. 215° (dec.).

The salt can be stored in an inert atmosphere without deterioration.

3. Tetrahydrofuran is dried by distillation from lithium aluminium hydride.

4. o-Methoxyphenylmagnesium bromide is prepared from 77.5 g. (0.414 mole, distilled from calcium hydride) of o-bromoanisole and 11 g. (0.46 g.-atom) of magnesium turnings in 100 ml. of dry tetrahydrofuran. The solution of this Grignard reagent is heated to reflux for 30 minutes prior to use.

5. Hexamethylphosphoramide is dried by distillation from calcium hydride.

6. If pure oxazoline is required, residual amounts of hexamethylphosphoramide can be removed by elution of the ethereal solution through silica gel (20–40 mesh).

7. Ice may be added to keep the mixture cool during the neutralization.

8. If this step is omitted, the o-anisaldehyde obtained after hydrolysis of the oxazolidine is contaminated with 5–10% o-bromoanisole.

9. o-Anisaldehyde is commercially available from Aldrich Chemical Co. and Eastman Organic Chemicals, Eastman Kodak Co.

3. Discussion

The conversion of a Grignard or an organolithium reagent to an aldehyde has been accomplished by a variety of reagents. The methods include such reagents as N-ethoxymethyleneaniline;[2a] ethyl orthoformate;[2a] p-dimethylaminobenzaldehyde;[2b] dimethyl formamide;[3] a dihydroquinazolinium salt;[4] and, more recently, a tert-alkyl isonitrile.[5]

TABLE I
ALDEHYDES FROM N,4,4-TRIMETHYL-2-OXAZOLINIUM IODIDE

Grignard Reagent	Aldehyde	Yield, %
C_6H_5MgBr	C_6H_5CHO	69
$C_6H_5CH_2MgCl$	$C_6H_5CH_2CHO$	87
$C_6H_5CH{=}CHMgBr$	$C_6H_5CH{=}CHCHO$	64
$C_6H_5C{\equiv}CMgBr$	$C_6H_5C{\equiv}CCHO$	51
$o\text{-}CH_3OC_6H_4MgBr$	$o\text{-}CH_3OC_6H_4CDO$	70[a]

[a] From 2-deuterio-N,4,4-trimethyl-2-oxazolinium iodide.

This procedure illustrates a general method for the preparation of aryl, benzyl, alkynyl, and vinyl aldehydes.[6] Table I gives the aldehydes which have been prepared from the corresponding Grignard reagents by conditions similar to those described here.

This method does not allow the formylation of aliphatic Grignard or organolithium reagents since in these cases the enhancement in base strength in the presence of hexamethylphosphoramide produces side reactions due to proton abstraction.

The present method is simple, proceeds easily and in good yield. The starting materials are readily available. The method is of particular value for the ready preparation of C-1 deuterated aldehydes using the 2-deuterio-N,4,4-trimethyl-2-oxazolinium iodide.[6] Also, since ^{14}C-labeled formic acid is routinely available, this provides easy access to isotopically labeled aldehydes.

1. Department of Chemistry, Colorado State University, Fort Collins, Colorado 80521.
2. (a) L. I. Smith and J. Nichols, *J. Org. Chem.*, **6**, 489 (1941); (b) A. J. Sisti, *Org. Syn.* **44**, 4 (1964).
3. R. A. Barnes and W. M. Bush, *J. Amer. Chem. Soc.*, **81**, 4705 (1959).
4. H. M. Fales, *J. Amer. Chem. Soc.*, **77**, 5118 (1955).
5. H. M. Walborsky, W. M. Morrison, III, and G. E. Niznik, *J. Amer. Chem. Soc.*, **92**, 6675 (1970).
6. A. I. Meyers and E. W. Collington, *J. Amer. Chem. Soc.*, **92**, 6676 (1970).

ALKYLATIONS OF ALDEHYDES via REACTION OF THE MAGNESIOENAMINE SALT OF AN ALDEHYDE: 2,2-DIMETHYL-3-PHENYLPROPIONALDEHYDE

(Propionaldehyde, 2,2-dimethyl-3-phenyl-)

$$(CH_3)_2CHCHO + (CH_3)_3CNH_2 \xrightarrow[\text{1 hour}]{K_2CO_3} (CH_3)_2CHCH{=}NC(CH_3)_3$$

$$(CH_3)_2CHCH{=}NC(CH_3)_3 + C_2H_5MgBr \xrightarrow[\text{12-14 hours}]{\text{tetrahydrofuran}}$$

$$\left[\begin{array}{c} (CH_3)_2C{=}CH{-}N{-}C(CH_3)_3 \\ | \\ MgBr \end{array} \right] + C..H_6$$

$$\left[\begin{array}{c} (CH_3)_2C{=}CHN{-}C(CH_3)_3 \\ | \\ MgBr \end{array} \right] + C_6H_5CH_2Cl \xrightarrow[\text{2. 10\% aq. HCl}]{\text{1. reflux, 20 hours}}$$

$$C_6H_5CH_2C(CH_3)_2CHO$$

Submitted by G. STORK[1] and S. R. DOWD
Checked by D. R. WILLIAMS and G. BÜCHI

1. Procedure

A. *N-(2-Methylpropylidene)-tert-butylamine.* A 100-ml., three-necked, round-bottom flask equipped with a condenser, a nitrogen inlet tube, a 50-ml. dropping funnel, and a magnetic stirring bar is evacuated through a mercury bubbler, flamed dry, and flushed with nitrogen three times. The flask is charged with 36.0 g. (0.5 mole) of *tert*-butylamine (Note 1), and 36.5 g. (0.5 mole) of isobutyraldehyde (Note 1) is placed in the dropping funnel. Half of the isobutyraldehyde is added slowly through the dropping funnel, and then the remaining half-volume is added rapidly. The milky solution is allowed to stand at room temperature for 1 hour; the water layer is then pipeted out, and excess anhydrous potassium carbonate is added. Filtration and then distillation of this reaction mixture gives 32.0 g. (50%) of N-(2-methylpropylidene)-*tert*-butylamine, b.p. 50° (75 mm.) (Note 2).

B. *2,2-Dimethyl-3-phenylpropanal.* A 100-ml., three-necked, round-bottom flask equipped with an ether condenser, a nitrogen inlet tube, a 50-ml. Herschberg dropping funnel, and a magnetic stirring bar is evacuated through a mercury bubbler, flamed dry, and flushed with nitrogen three times. The system is left under a slight positive pressure of nitrogen, and all the reactants are added under a stream of nitrogen. A solution of 0.05 mole of ethylmagnesium bromide in 37 ml. of tetrahydrofuran (Note 3) is placed in the flask. A solution of 6.35 g. (0.05 mole) of N-(2-methylpropylidene)-*tert*-butylamine (Note 4) in 5 ml. of tetrahydrofuran (Note 5) is then added from the dropping funnel. The resulting mixture is refluxed for 12–14 hours until 1 mole-equivalent of gas is evolved (Note 6). The reaction mixture is then cooled to room temperature, and 6.30 g. (0.05 mole) of benzyl chloride (Note 1) is added from the dropping funnel. The solution is then refluxed for 20 hours, at which time the pH is 9–10 (pHydrion paper). To the cooled solution, which contains a large amount of solid, is added 20–30 ml. of 10% hydrochloric acid solution. The clear, yellow-brown solution is then refluxed for 2 hours. The cooled solution is saturated with solid sodium chloride and extracted five times with diethyl ether. The organic extracts are washed once with 25 ml. of 5% hydrochloric acid solution, and then repeatedly with brine until the washings are neutral. The organic layer is dried over anhydrous magnesium sulfate, filtered, and the solvent is removed at atmospheric pressure through a 12-in. Vigreux column fitted with a partial take-off head. Distillation of the residue (Note 7) through a 20-in. vacuum-jacketed fractionating column affords 5.1–5.4 g. (63–66%) of 2,2-dimethyl-3-phenylpropanal, b.p. 70–73° (1.5 mm.) (Note 8).

2. Notes

1. These reagents were obtained from Eastman Organic Chemicals and were not distilled prior to use.

2. The checkers found that the yield could be improved (37–38 g., 58–60%) if the reaction mixture was allowed to remain over anhydrous potassium carbonate for 8–12 hours.

3. The ethylmagnesium bromide is prepared in dry tetrahydrofuran and stored no longer than 1 week in a 250-ml. tube fitted with a 3-way vacuum stopcock and a dropping buret. The solution is decanted into the buret, and the correct volume is transferred to the reaction flask with positive nitrogen pressure. The tetrahydrofuran is purified by distillation from lithium aluminium hydride. See *Org. Syn.*, **46,** 105 (1966), for warning regarding the purification of tetrahydrofuran.

4. The aldimine is freshly distilled [b.p. 50° (75 mm.)] prior to use.

5. A vigorous reaction may result. At this stage of the reaction, control is maintained with an ice–water bath.

6. The volume of gas evolved is estimated with an inverted cylinder filled with water attached by rubber tubing to the outlet of the mercury bubbler.

7. Gas chromatographic analysis at 155° on a 5 ft. 5% SE-30 column shows only the presence of 2,2-dimethyl-3-phenylpropanal and solvent.

8. The infrared spectrum (CHCl$_3$) showed absorption at 2705, 1725, and 1605 cm^{-1}. The 2,4-dinitrophenylhydrazone, recrystallized from ethanol–ethyl acetate as long, yellow-orange needles, melts at 150–152° (reported[3] 154–155°).

3. Discussion

This procedure illustrates the mono-alkylation of α-substituted aldehydes by means of the metalloenamine method.[4] The preparation of the aldimine has been adapted from the procedure of R. Tiollais.[2] The procedure is useful in the preparation of aldimines from low-boiling components. The readily prepared aldimine is treated with an alkyl Grignard reagent generating the magnesioenamine halide salt, which can be alkylated with a variety of alkylating agents at the α-position. The yields are high, monoalkylation is the exclusive reaction, and there is no rearrangement when using an allylic halide. The general method is applicable to the alkylation of ketones via the magnesium bromide salts of the corresponding ketimine (Table I).

TABLE I
REACTION OF VARIOUS ALDIMINE AND KETIMINE MAGNESIUM BROMIDE SALTS WITH ALKYLATING AGENTS IN TETRAHYDROFURAN

Imine	Halide	Yield, %
N-(2-methylpropylidene)-tert-butylamine	n-butyl iodide	65
N-(heptylidene)-tert-butylamine	n-butyl iodide	47
N-(propylidene)-tert-butylamine	n-butyl bromide	60
N-(cyclohexylidene) cyclohexylamine	n-butyl iodide	78
	isopropyl iodide	61
	benzyl chloride	60
N-(cyclopentylidene) cyclohexylamine	n-butyl iodide	72
N-(cycloheptylidene) cyclohexylamine	n-butyl iodide	75

In a variation of this method, metalloenamines have been generated from the aldimine and lithium diisopropylamide in ether and have been alkylated in a limited number of cases.[5] A further variation, using lithium dialkylamides in hexamethylphosphoramide, has been shown by Th. Cuvigny and H. Normant[6] to give good yields of alkylated aldehydes. However, secondary alkyl halides fail to react, giving the dehydrohalogenation product instead. Another approach is that of Meyers and co-workers,[7] which involves the alkylation of the lithio salt of 2-methyldihydro-1,3-oxazines. It suffers from the necessity of carrying out several low-temperature steps and a pH-controlled borohydride reduction.

1. Department of Chemistry, Columbia University, New York, New York 10027.
2. R. Tiollais, *Bull. Soc. Chim. Fr.*, **14**, 708 (1947).
3. G. Opitz, H. Heilmann, H. Mildenberger, and H. Suhr, *Justus Liebigs Ann. Chem*, **649**, 36 (1961).
4. G. Stork and S R. Dowd, *J. Amer. Chem. Soc.*, **85**, 2178 (1963).
5. G. Wittig, H.-D. Frommeld, and P. Suchanek, *Angew. Chem. Int. Ed. Engl.*, **2**, 683 (1963); G. Wittig and H.-D. Frommeld, *Chem. Ber.*, **97**, 3548 (1964).
6. Th. Cuvigny and H. Normant, *Bull. Soc. Chim. Fr.*, 3976 (1970).
7. A. I. Meyers, A. Nabeya, H. W. Adickes, and I. R. Politzer, *J. Amer. Chem. Soc.*, **91**, 763 (1969).

DIRECTED ALDOL CONDENSATIONS: *THREO*-4-HYDROXY-3-PHENYL-2-HEPTANONE

(2-Heptanone, 4-Hydroxy-3-phenyl)

$$C_6H_5CH_2COCH_3 \xrightarrow[\text{2. }(CH_3CO)_2O,\, 5\text{-}30°]{\text{1. NaH, }CH_3OCH_2CH_2OCH_3,\, 20°}$$

(C₆H₅)(H)C=C(OCOCH₃)(CH₃)

(C₆H₅)(H)C=C(OCOCH₃)(CH₃)

$$\xrightarrow[\substack{2)\ ZnCl_2,\ (C_2H_5)_2O \\ 3)\ n\text{-}C_3H_7CHO,\ 0\text{-}10° \\ \overline{N}H_4Cl,\ H_2O,\ 0°}]{\substack{1)\ CH_3Li\ (2\ equiv.) \\ CH_3OCH_2CH_2OCH_3}}$$

CH₃COC(C₆H₅)(H)—C(OH)(H)(CH₂)₂CH₃

Submitted by ROBERT A. AUERBACH, DAVID S. CRUMRINE, DAVID L. ELLISON, and HERBERT O. HOUSE[1]
Checked by WATARU NAGATA and NORBUHIRO HAGA

1. Procedure

Caution! Since hydrogen is liberated, this preparation should be performed in a hood.

A. *2-Acetoxy-trans-1-phenylpropene.* A dry, 500-ml., three-necked flask is fitted with a mechanical stirrer, a pressure-equalizing dropping funnel, and a rubber septum, and the apparatus is arranged so that the flask may be cooled intermittently with an ice bath. After the reaction vessel has been flushed with nitrogen (admitted through a hypodermic needle in the rubber septum) a static nitrogen atmosphere is maintained in the reaction vessel for the remainder of the reaction. Into the flask is placed 35 g. of a 57% dispersion of sodium hydride (20 g. or 0.83 mole) in mineral oil (Note 1). The mineral oil is washed from the sodium hydride with a 200-ml. portion of anhydrous pentane. The supernatant pentane layer is removed by means of a stainless steel cannula inserted through the rubber septum (Note 2). The residual sodium hydride is mixed with 250 ml. of anhydrous 1,2-dimethoxyethane (Note 3), and then 65 g. (0.48 mole) of phenylacetone (Note 4) is added dropwise and with stirring during 50–60 minutes. During this addition an open hypodermic needle should be inserted in the rubber septum to permit the escape of hydrogen and intermittent cooling with an ice bath may be necessary to keep the reaction solution from boiling. The resulting mixture is stirred for 3 hours while it is allowed to cool, and then the mixture is allowed stand for approximately 2 hours to permit the excess sodium hydride to settle. The supernatant liquid is transferred under positive nitrogen pressure through a stainless steel cannula (Note 2) into a 1-l., three-necked flask containing 108 g. (100 ml. or 1.00 mole) of cold (0°), freshly distilled acetic anhydride (b.p. 140°) and fitted with a mechanical stirrer, a thermometer, an ice bath, and a rubber septum into which are inserted a hypodermic needle to admit nitrogen and the cannula used to transfer the enolate solution.

The enolate solution is added slowly with cooling and vigorous stirring so that the temperature of the reaction mixture remains below 30°. After all the supernatant enolate solution has been transferred, the residual slurry of sodium hydride is washed with an additional 50-ml. portion of anhydrous 1,2-dimethoxyethane (Note 3), and these washings are also added to the acetic anhydride solution. The resulting viscous mixture is stirred at room temperature for an additional 30 minutes and then poured cautiously into a mixture of 500 ml. of pentane, 500 ml. of water, and 130 g. (1.54 moles) of sodium bicarbonate. When hydrolysis of the excess acetic anhydride and neutralization of the acetic acid are complete, the pentane layer is separated, and the aqueous phase is extracted with 100 ml. of pentane. The combined pentane solutions are dried over anhydrous magnesium sulfate and then concentrated under reduced pressure with a rotary evaporator. Distillation of the residual orange liquid through a 20–30-cm. Vigreux column (Note 5) provides 61.7–80.6 g. (73–95%) of the 2-acetoxy-*trans*-1-phenylpropene, b.p. 82–89° (1 mm.), n^{25}D 1.5320–1.5327 (Note 6).

B. *threo-4-Hydroxy-3-phenyl-2-heptanone*. A dry, 500-ml., three-necked flask is fitted with a Teflon®-coated magnetic stirring bar, a gas inlet tube equipped with a stopcock, a low-temperature thermometer, and a rubber septum and is mounted to permit the use of an external cooling bath. The apparatus is flushed with nitrogen, and a static nitrogen atmosphere is maintained in the reaction vessel throughout the reaction. After 10–20 mg. of 2,2'-bipyridyl has been added to the reaction flask as an indicator, an ethereal solution containing 0.412 mole of halide-free methyllithium (Note 7) is added to the reaction flask with a hypodermic syringe or stainless steel cannula inserted through the rubber septum. The ether is removed under reduced pressure (Note 8) while the flask is warmed to 40° with a water bath, and then the reaction vessel is refilled with nitrogen, and 120 ml. of anhydrous 1,2-dimethoxyethane is added (Note 3). The resulting purple solution is cooled to −10 to −20° with a cooling bath containing dry ice and isopropyl alcohol, and the 35.2 g. (0.200 mole) (Note 9) of 2-acetoxy-*trans*-1-phenylpropene is added from a hypodermic syringe dropwise and with stirring during 15 minutes while the temperature of the reaction mixture is kept in the range −20 to +10°. The resulting red-brown solution is stirred for an additional 10 minutes at −10 to 0°, and then 285 ml. of an etheral solution containing 0.202 mole of anhydrous zinc chloride (Note 10) is added to the cold (−10 to +10°) reaction mixture from a hypodermic syringe dropwise and with stirring during 10 minutes. The resulting reddish-yellow cloudy reaction mixture (Note 11) is stirred at 0° for 10 minutes, and then 14.50 g. (0.201 mole of freshly distilled butyraldehyde (Note 12) is added rapidly (30 seconds) and with stirring to the cold (−5 to +10°) reaction mixture. After the resulting mixture has been stirred at 0–5° for 4 minutes, it is poured with vigorous stirring into a cold (0–5°) mixture of 500 ml. of aqueous 4M ammonium chloride and 200 ml. of ether. The ether layer is separated, and the aqueous phase is extracted with two 200-ml. portions of ether. The combined organic solutions are washed successively with two 100-ml. portions of aqueous $1 M$ ammonium chloride and with two 50-ml. portions of saturated aqueous sodium chloride, and these combined aqueous washings are extracted with an additional 100-ml. portion of ether. The combined ether solutions are dried over anhydrous magnesium sulfate and then concentrated under reduced pressure (water aspirator) with a rotary evaporator to remove the solvents, including the residual 1,2-dimethoxyethane. The residual liquid, which may crystallize on standing (Note 13), is triturated with 50 ml. of pentane, and the crystalline solid that separates is collected on a filter. The filtrate is concentrated under reduced pressure and again triturated with pentane to separate an additional crop of the crude product. The combined crops of the crude *threo*-aldol product amount to 26.2–28.4 g. (64–69%), m.p. 57–62°. The crude product is recrystallized from 125–150 ml. of hexane. After the solution has been cooled to 0°, 21.3–24.1 g. of the *threo*-aldol product is collected as white needles, m.p. 71.5–72.5° (Note 14). The mother liquors are concentrated and cooled to separate additional fractions of the product (0.5–0.8 g.), m.p. 71–72°. The total yield of the *threo*-aldol product is 22.1–24.6 g. (53–60%).

2. Notes

1. The submitters used a 57% dispersion of sodium hydride in mineral oil obtained from Alfa Inorganics, Inc., and the checkers used a 50% dispersion of sodium hydride in mineral oil obtained from Metal Hydrides, Inc.

2. As a stainless steel cannula was not available, the checkers made a minor modification in the operation without any trouble. They transferred the supernatant pentane and the solution of the sodium enolate using a Luer-lock hypodermic syringe with a stainless steel needle, preflushed with nitrogen, and the apparatus was swept with nitrogen during this operation.

3. The submitters distilled 1,2-dimethoxyethane (b.p. 83°) from lithium alminium hydride immediately before use. The checkers distilled from sodium hydride immediately before use.

4. The submitters used a commercial sample of phenylacetone, obtained from Aldrich Chemical Company, Inc.; the checkers used

material of the same grade obtained from Maruwaka Chemical Industries Ltd. (Japan). It was used without further purification.

5. The checkers used a 15 × 1 cm., unpacked, vacuum-jacketed column instead of Vigreux column for the distillation.

6. The results of gas chromatographic analysis of the products made by the submitters are as follows: On a 3-m. gas chromatography column, packed with silicone fluid QF_1 supported on Chromosorb P, and heated to 190°, the product exhibits peaks at 5.8 minutes corresponding to 2–3% phenylacetone, at 7.5 minutes corresponding to 97–98% of the enol acetate (*cis* and *trans* isomers not resolved), and at 8.0 minutes corresponding to a trace (<1%) of 3-phenyl-2,4-pentanedione. On a second, 7-m. gas chromatography column, packed with silicone fluid DC-710 on Chromosorb P and heated to 190°, the product exhibits peaks at 21.0 minutes corresponding to phenylacetone, at 39.0 minutes corresponding to the *trans*-enol acetate (97–98% of the product), and at 42.2 minutes corresponding to the *cis*-enol acetate (2–3% of the product). The checkers used a 45 m. × 0.25 mm. stainless steel column (Golay type) coated with Apiezon L, which was heated at 150° and swept with helium at 1.5 kg./cm.^{2} The product exhibits peaks at 5.5 minutes corresponding to phenylacetone (2–3% of product), at 14.2 minutes corresponding to the *trans*-enol acetate (91–92% of the product), and at 15.8 minutes corresponding to the *cis*-enol acetate (5–6% of the product). The product has infrared absorption (CCl_4 solution) at 1765 (enol ester C=O) and 1685 cm^{-1} (enol ester C=C) with ultraviolet maxima (95% EtOH solution) at 248.5 mμ (ϵ 18,000) and 325 mμ (ϵ 415) and proton magnetic resonance peaks (CCl_4 solution) at 7.0–7.4 (5, multiplet, aryl C*H*), 5.82 (1, partially resolved multiplet, vinyl C*H*), and 2.01 (6, partially resolved multiplet, CH_3CO and vinyl C*H*$_3$). The mass spectrum of the product has a parent ion at *m/e* 176 with abundant fragment peaks at *m/e* 134, 91, 45, 43, and 39.

7. The submitters used an ether solution of halide-free methyllithium, purchased from Foote Mineral Company, and the checkers prepared the compound from methyl chloride and lithium metal in ether according to the literature.[2] The solution was standardized before use by the titration procedure described in a previous volume of this series.[3] The checkers observed that use of a halide-containing ether solution of methyllithium resulted in a considerable decrease in yield of the product, principally due to difficulty in following the subsequent procedure described in the text.

8. A convenient apparatus for evacuating the reaction vessel and then refilling it with nitrogen is described in an earlier volume of this series.[3]

9. If the violet color of the reaction solution is completely discharged, indicating that all the methyllithium has been consumed, addition of the enol acetate should be stopped at that point. The actual concentration of enolate anion in the solution can be calculated from the amount of enol acetate added.

10. To prepare an ethereal solution of anhydrous zinc chloride (m.p. 283°), the submitters placed 50.0 g. (0.369 mole) of pulverized reagent zinc chloride, obtained from either Mallinckrodt Chemical Works or Fisher Scientific Company, in a 1-l., round-bottomed flask, and the vessel was evacuated to about 1 mm. pressure. The flask was heated strongly with a burner with swirling until as much of the solid as practical had been melted. The evacuated flask was cooled and shaken (*Caution! Perform this operation behind a safety shield in a hood and with heavy gloves to protect the operator's hands in case the flask should implode*) to break up the large lumps of zinc chloride. This fusion under reduced pressure should be repeated three times. To the resulting anhydrous zinc chloride was added 500 ml. of anhydrous ether that had been freshly distilled from lithium aluminium hydride. The resulting mixture was refluxed for 3 hours under a static nitrogen atmosphere, and then the mixture was allowed to stand until the undissolved solid had settled. The resulting supernatant solution was transferred with a stainless steel cannula under positive nitrogen pressure (Note 2) into a second dry flask or Schlenk tube capped with a rubber septum. Aliquots of this solution, diluted with aqueous ammonia, can be titrated with standard EDTA solution to a Erichrome Black T endpoint to determine the zinc content.[4] Alternatively, the chloride ion concentration of aliquots can be determined by a Volhard titration. Typical values found for these ether solutions are $0.73 M$ in zinc ion and $1.38 M$ in chloride ion. Thus, this solution is 0.69–$0.73 M$ in zinc chloride. If the final solution is significantly less concentrated than $0.7 M$ in zinc chloride, it is probable that the dehydration of the solid zinc chloride was not complete. In this event, the submitters recommend that a fresh solution of zinc chloride be prepared with more attention to the initial dehydration of the solid zinc chloride. The checkers used pulverized reagent zinc chloride, obtained from Wako Pure Chemical Industries Ltd. (Japan).

11. The white precipitate that separates is a part of the lithium

chloride formed in the reaction mixture. Separation of the material is not necessary.

12. The submitters used a commercial grade of butyraldehyde from Eastman Organic Chemicals; the checkers used the butyraldehyde of the same grade from Wako Pure Chemical Industries Ltd. (Japan) and distilled it before use, b.p. 72–74°.

13. The proton magnetic resonance spectrum of a C_6D_6 solution of the crude product exhibits benzylic CH doublets at δ 3.58 (J = 9.4 Hz., attributable to 90–96% of the *threo* aldol isomer) and at δ 3.42 (J = 5.3 Hz., attributable to 4–10% of the *erthro* aldol isomer) downfield from internal tetramethylsilane. This mixture may be separated by chromatography on acid-washed silicic acid to permit the isolation of both the *threo* and the *erythro* diastereoisomers.[6]

14. The *threo*-hydroxy ketone exhibits infrared absorption (CCl_4 solution) at 3540 (associated OH) and 1705 cm^{-1} (C=O) with a series of weak (ϵ 300 or less) ultraviolet maxima (95% EtOH solution) in the region 240–270 mμ as well as a maximum at 286 mμ (ϵ 345). The proton magnetic resonance spectrum (CCl_4 solution) of the product shows resonance at 7.1–7.5 (5H, multiplet, aryl CH), 4.0–4.4 (1H, multiplet, CH—O), 3.65 (1H, doublet, J = 9.5 Hz., benzylic CH), 3.35 (1H, singlet, OH), 2.03 (3H, singlet, CH_3CO), and 0.6–1.9 p.p.m. (7H, multiplet, aliphatic CH) downfield from internal tetramethylsilane. The mass spectrum of the product exhibits the following relatively abundant peaks: *m/e* (relative intensity), 206 (M$^+$, 0.1), 188 (8), 146 (20, 135 (26), 134 (100), 117 (52), 92 (48), 91 (76), 65 (31), 44 (36), and 43 (60).

3. Discussion

The present procedures illustrate general methods for the use of preformed lithium enolates[5] as reactants in the aldol condensation[6] and for the quenching of alkali metal enolates in acetic anhydride to form enol acetates with the same structure and stereochemistry as the starting metal enolate.[7] The aldol product, *threo*-4-hydroxy-3-phenyl-2-heptanone, has been prepared only by this procedure.

The methods previously used to obtain single aldol products (or their dehydrated derivatives) from reactants where several aldol products are possible[8] include the reaction of bromozinc enolates, from α-bromoketones, with aldehydes;[9] the reaction of bromomagnesium enolates, from either α-bromoketones or from ketones and bromomagnesium amides or sterically hindered Grignard reagents, with aldehydes;[10,11] and the reaction of α-lithio derivatives of imines with aldehydes or ketones.[12] Like the present procedure, each of these methods relies upon trapping the intermediate β-keto alkoxide derivative as a metal chelate in an aprotic reaction solvent. The present procedure increases the versatility of the aldol condensation by utilizing the variety of specific lithium enolates that can be generated from unsymmetrical ketones.[5]

TABLE I[6]
DIRECTED ALDOL CONDENSATIONS WITH PREFORMED LITHIUM ENOLATES IN THE PRESENCE OF ZINC CHLORIDE

Enolate	Aldol Product	Yield, %	Stereoisomer ratio, threo/erythro
$(CH_3)_3C-C(O^-Li^+)=CH_2$	$(CH_3)_3CCOCH_2CH(OH)C(CH_3)_3$	82	—
1-lithio-oxy-cyclohexene	2-(α-hydroxybenzyl)cyclohexanone	76	4/1
$C_6H_5CH=C(O^-Li^+)-CH_3$	$C_6H_5CH(OH)-CH(COCH_3)-C_3H_7\text{-}n$	80	9/1
2-methyl-6-tert-butyl-1-lithiooxycyclohexene	4-tert-butyl-2-methyl-2-(α-hydroxybenzyl)cyclohexanone	84	2/1[a]
$n\text{-}C_4H_9CH=C(O^-Li^+)-CH_3$	$n\text{-}C_4H_9CH(OH)-CH(COCH_3)-C_6H_5$	80	1/1

[a] The aldol product contained 70% of isomers with an axial α-hydroxybenzyl substituent.

In this procedure the lithium enolate is treated successively with anhydrous zinc chloride and with an aldehyde to form the zinc(II) chelate of a β-keto alkoxide. The optimum quantity of zinc chloride is that amount required to form zinc(II) salts of all strong bases in the reaction mixture. Thus, 1 mole of zinc chloride should be added for each mole of lithium enolate (and accompanying lithium t-butoxide) formed from an enol acetate as in the present example. If the lithium enolate is formed from the ketone and lithium diisopropylamide or from a trimethylsilyl enol ether and methyllithium, then 0.5 mole of zinc chloride should be used for each mole of lithium enolate. The optimum reaction solvent is either ether or ether–1,2-dimethoxyethane mixtures with a reaction temperature of -10 to $+10°$ and a reaction time of 2–5 minutes. Longer reaction times and higher reaction temperatures may lead to a variety of by-products resulting from polycondensation and dehydration. The aldol products are efficiently isolated by *adding the reaction mixtures* to a cold (0–5°), aqueous solution of ammonium chloride followed by rapid separation of the aldol products. Since many of the aldol products are especially prone to epimerization, dehydration, or reversal of the aldol condensation, they should not be exposed to strong acids or strong bases. Mixtures of stereoisomeric aldol products with similar physical properties can usually be separated by chromatography on acid-washed silicic acid.[6,13]

In several cases (including the present example) where diastereoisomeric aldol products are possible, there is a preference for the formation of the *threo*-diastereoisomer. This stereochemical preference presumably arises because the six-membered cyclic zinc chelate of the *threo*-isomer can exist in a chair conformation with both substituents in equatorial positions. Table I summarizes the results obtained in several aldol condensations performed by the present procedure.

1. School of Chemistry, Georgia Institute of Technology, Atlanta, Georgia 30332.
2. H. J. Berthold, G. Groh, and K. Stoll, *Z. Anorg. Allg. Chem.*, **37**, 53 (1969).
3. M. Gall and H. O. House, *Org. Syn.*, **52**, 39 (1972).
4. W. Biederman and G. Schwarzenbach, *Chimia*, **2**, 56 (1948); G. Schwarzenbach and H. Flaschka, "Complexometric Titrations," 2nd. ed., Methuen and Company, London, 1969, p. 260.
5. For examples and leading references, see H. O. House, M. Gall, and H. D. Olmstead, *J. Org. Chem.*, **36**, 2361 (1971).
6. H. O. House, D. S. Crumrine, A. Y. Teransishi, and H. D. Olmstead, *J. Amer. Chem. Soc.*, **95**, 3310 (1973).
7. H. O. House, R. A. Auerbach, M. Gall, and H. D. Olmstead, *J. Org. Chem.*, **38**, 514 (1973).

8. For a general review of the aldol condensation, see A. T. Nielsen and W. J. Houlihan, *Org. Reactions*, **16**, 1 (1968).
9. T. A. Spencer, R. W. Britton, and D. S. Watt, *J. Amer. Chem. Soc.*, **89**, 5727 (1967).
10. A. T. Nielsen, C. Gibbons, and C. A. Zimmerman, *J. Amer. Chem. Soc.*, **73**, 4696 (1951).
11. J. E. Dubois and P. Fellmann, *Compt. Rend. Ser. C*, **274**, 1307 (1972).
12. G. Wittig and A. Hesse, *Org. Syn.*, **50**, 66 (1970); G. Wittig and H. Rieff, *Angew. Chem. Int. Ed. Engl.* **7**, 7 (1968).
13. H. Brockmann and H. Muxfeldt, *Chem. Ber.*, **89**, 1379 (1956).

1-BENZYLINDOLE

(Indole, 1-benzyl)

$$\text{Indole} \xrightarrow[\text{DMSO, 25°}]{\text{KOH, C}_6\text{H}_5\text{CH}_2\text{Br}} \text{1-benzylindole}$$

Submitted by HARRY HEANEY and STEVEN V. LEY[1]
Checked by A. BROSSI, E. E. GARCIA, and R. P. SCHWARTZ

1. Procedure

A 500-ml. Erlenmeyer flask equipped with a magnetic stirring bar is charged with 200 ml. of dimethylsulfoxide (DMSO) (Note 1) and 26.0 g. (0.4 mole) of potassium hydroxide (Note 2). The mixture is stirred at room temperature for 5 minutes, and then 11.7 g. (0.1 mole) of indole (Note 3) is added. The stirring is continued for 45 minutes before 34.2 g. (0.2 mole) of benzyl bromide (Note 4) is added (Note 5). After stirring for an additional 45 minutes the mixture is diluted with 200 ml. of water. The mixture is extracted with three 100-ml. portions of ether, and each ether layer is washed with three 50-ml. portions of water. The combined ether layers are dried over calcium chloride, and the solvent is removed at slightly reduced pressure (Note 6). The excess benzyl bromide is removed by distillation at approximately 15 mm., and the residue is distilled to separate 17.5–18.4 g. (85–89%) of 1-benzylindole, b.p. 133–138° (0.3 mm.). The distillate crystallizes upon cooling and scratching. Recrystallization from ethanol gives material melting at 42–43° (Notes 7 and 8).

2. Notes

1. The dimethylsulfoxide used was not rigorously dried but should not contain an appreciable amount of water.

2. Freshly crushed 86% potassium hydroxide pellets were used.

3. A commercial grade of indole is satisfactory.

4. Reagent grade benzyl bromide was used without further purification.

5. Cooling with an ice–water bath moderates the exothermic reaction.

6. The submitters used a Büchi rotary evaporator (water aspirator).

7. The submitters have obtained yields as high as 20 g. (97%).

8. The recrystallized product exhibits a proton magnetic resonance spectrum (CDCl$_3$) δ 5.21 (singlet, $2H$), 6.52 (doublet, $J = 3.4$ Hz., $1H$), 7.0–7.4 (multiplet, $9H$), and 7.5–7.7 (multiplet, $1H$).

3. Discussion

Although the N-alkylation of pyrrole[2] and indole[3] has been reported on many occasions, a generally applicable, simple, high-yield procedure was not available. Many simple procedures give mixtures of products because of the ambident nature of the anions. However, alkylation at nitrogen is usually predominant in strongly ionizing solvents. Recent methods include alkylations of indole in liquid ammonia,[4] dimethylformamide,[5] and hexamethylphosphoramide.[6]

The use of dimethyl sulphoxide as a dipolar aprotic solvent is well known,[7] and the present method can be regarded as a model procedure and has been applied to the preparation of a number of N-n-alkylpyrroles and N-n-alkyl indoles.[8] The yield of N-benzylindole is considerably higher than in previously reported preparations and is as good as that reported for the preparation of N-methylindole in liquid ammonia.[4] The present method is, however, less laborious and quicker to carry out. Very high yields are obtained in reactions using n-alkyl halides and moderately good yields with secondary alkyl halides. The reactions should be compared with those recently reported for pyrryl-thallium.[9]

1. Department of Chemistry, The University of Technology, Loughborough, Leicestershire, LE11 3TU, England.
2. K. Schofield, "Heteroaromatic Nitrogen Compounds, Pyrroles and Pyridines," Butterworths, London, 1967.
3. L. R. Smith, Indoles, Part Two, in W. J. Houlihan, "The Chemistry of Heterocyclic Compounds," Wiley-Interscience, New York, 1972.

4. K. T. Potts and J. E. Saxton, *Org. Syn.*, **40**, 68 (1960).
5. J. Szmuzkovicz, *Belg. Pat.*, 621,047 (1963).
6. H. Normant and T. Cuvigny, *Bull. Soc. Chim. Fr.*, 1866 (1965).
7. L. F. Fieser and M. Fieser, "Reagents for Organic Synthesis," John Wiley, New York, Vol. 1, 1967, pp. 296–318; Vol. 2, 1969, pp. 157–173; Vol. 3, 1972, pp. 119–123; D. J. Cram, B. Rickborn, and G. R. Knox, *J. Amer. Chem. Soc.*, **82**, 6412 (1960); D. J. Cram, B. Rickborn, C. A. Kingsbury, and P. Haberfield, *J. Amer. Chem. Soc.*, **83**, 3678 (1961).
8. H. Heaney and S. V. Ley, *J. Chem. Soc. Perkin 1*, 499 (1973).
9. C. F. Candy and R. A. Jones, *J. Org. Chem.*, **36**, 3993 (1971).

N-ALKYLINDOLES FROM THE ALKYLATION OF INDOLE SODIUM IN HEXAMETHYLPHOSPHORAMIDE: 1-BENZYLINDOLE

(Indole, 1-benzyl)

$$2 \; \text{indole} \xrightarrow[5 \text{ hrs., } 25°]{\text{NaH, HMPA}} 2 \; [\text{indole-N}^\ominus \text{Na}^\oplus] + H_2$$

$$[\text{indole-N}^\ominus \text{Na}^\oplus] \xrightarrow[0-25°, 8-15 \text{ hrs.}]{C_6H_5CH_2Cl} \text{1-benzylindole (N-CH}_2C_6H_5)$$

Submitted by GEORGE M. RUBOTTOM[1] and JOHN C. CHABALA[2]
Checked by R. E. IRELAND and JAMES E. KLECKNER

1. Procedure

A. *Indole Sodium.* In a 100-ml., three-necked flask fitted with a reflux condenser, a magnetic stirring bar, and a gas inlet tube is placed 2.34 g. (0.02 mole) of indole (Note 1) and 15 ml. of hexamethylphosphoramide (HMPA) (Note 2) under a static atmosphere of argon. The flask is cooled to 0° by means of an ice bath, and 0.53 g. (0.022 mole) of sodium hydride is added to the stirred solution over a period of 10 minutes (Note 3). The resulting slurry is then stirred for 5 hours at room temperature (Note 4).

B. *1-Benzylindole.* The slurry of indole sodium is cooled to 0° (ice bath), and 2.30 ml. (2.53 g., 0.02 mole) of benzyl chloride (Note 5) is added as rapidly as possible to the stirred mixture. The mixture is then stirred for 8–15 hours (overnight), during which time the ice in the ice bath melts and the temperature of the reaction flask gradually rises to room temperature. The mixture is then diluted with 15 ml. of water and extracted with three 25-ml. portions of ether. The combined ethereal extracts are washed with two 40-ml. portions of water and dried with anhydrous magnesium sulfate. After filtration the solvent is removed at reduced pressure, and 4.4 g. of crude 1-benzylindole is obtained as a liquid. After bulb-to-bulb distillation of this material in a Kügelrohr oven [120–130° (0.0025 mm.)], crystallization of the distillate from 15 ml. of hot ethanol affords 3.46–3.61 g. (83–87%) of 1-benzylindole. A second crop amounting to 0.17–0.26 g. (4–6%) is obtained on concentration of the mother liquors to 6 ml. The total yield of 1-benzylindole, m.p. 43–44°, is 3.72–3.78 g. (90–91%) (Notes 6 and 7).

2. Notes

1. Commercial indole (Matheson, Coleman and Bell) was used with no further purification.

2. Commercial HMPA (Aldrich) was stored over Linde 4 A Molecular Sieves and used without further purification.

3. A batch of 0.93 g. of a 57% sodium hydride dispersion in mineral oil is washed with hexane to remove the mineral oil immediately prior to use. The slow addition in the cold minimizes the small amount of foaming caused by hydrogen evolution.

4. This stirring time insures complete formation of the indole sodium.

5. Commercial benzyl chloride (Matheson, Coleman and Bell) is used without further purification.

6. The recrystallized product exhibits proton magnetic resonance peaks $(CDCl_3)\delta$: 5.21 ($2H$, singlet), 6.52 ($1H$, doublet, $J = 3.4$ Hz.), 7.0–7.4 ($9H$, multiplet), and 7.5–7.7 ($1H$, multiplet).

7. 1-Benzylindole colors significantly in contact with air at room temperature (*ca.* 1 week) but keeps indefinitely under argon.

3. Discussion

The alkylation of indole sodium generated from indole and sodium amide in liquid ammonia has been used as the general method for the

TABLE I
Alkylation of Indole Sodium with R-X in HMPA Solvent

R-X	Yield of N-Alkylindole, %	b.p. or m.p.
CH_3I	94	b.p. 73–75°/2.4 mm. (Ref. 14, b.p. 70–75/2 mm.)
C_2H_5I	92	b.p. 83–86°/0.6 mm. (Ref. 15, b.p. 82–85°/0.7 mm.)
$CH_2\!\!=\!\!CHCH_2Br$	84	b.p. 72–73°/0.12 mm. (Ref. 6, b.p. 114–116°/6 mm.)
$C_6H_5CH_2Cl$	90–91	m.p. 43–44° (Ref. 5, m.p. 44°)

preparation of N-alkylindoles.[3–11] The drawback to this method of synthesis is the necessity of using liquid ammonia. The procedure outlined here[12] overcomes this problem and yet affords pure N-alkylindoles in excellent yield. Further, the use of the HMPA–NaH base system affords conditions leading to the formation of the N-alkylindoles with little or no side reaction leading to C-alkylated products.[12,13] Table I illustrates the generality of the method.

1. Department of Chemistry, University of Puerto Rico, Rio Piedras, Puerto Rico 00931.
2. Department of Chemistry, Bucknell University, Lewisberg, Pennsylvania 17387.
3. K. T. Potts and J. E. Saxton, *J. Chem. Soc.*, 2641 (1954).
4. K. T. Potts and J. E. Saxton, *Org. Syn.*, **40**, 68 (1960).
5. H. Plieninger, *Chem. Ber.*, **87**, 127 (1954).
6. M. Nakazaki and S. Isoe, *Nippon Kagaku Zasshi*, **76**, 1159 (1955); *Chem. Abst.*, **51**, 17877.
7. M. Nakazaki, *Bull. Chem. Soc. Jap.*, **32**, 838 (1959).
8. N. I. Grineva, V. L. Sadovskaya and V. N. Ufimtsev, *Zh. Obshch. Khim.*, **33**, 552 (1963).
9. M. Julia and P. Manoury, *Bull. Soc. Chim. Fr.*, 1946 (1964).
10. S. Yamada, *Chem. Pharm. Bull.* (Tokyo), **13**, 88 (1965).
11. A. H. Jackson and A. E. Smith, *Tetrahedron*, **21**, 989 (1965).
12. Essentially the procedure outlined in G. M. Rubottom and J. C. Chabala, *Synthesis*, 566 (1972).
13. B. Cardillo, G. Casnati, A. Pochini, and R. Ricca, *Tetrahedron*, **23**, 3771 (1967).
14. L. Marion and C. W. Oldfield, *Can. J. Res. B*, **25**, 1 (1947).
15. A. P. Gray, H. Kraus, and D. E. Heitmeier, *J. Org. Chem.*, **25**, 1939 (1960).

GERANYL CHLORIDE

[(E)-1-Chloro-3,7-dimethyl-2,6-octadiene]

$$\text{Geraniol} + (C_6H_5)_3P + CCl_4 \xrightarrow[\text{reflux}]{CCl_4} \text{Geranyl chloride}$$

Submitted by JOSE G. CALZADA and JOHN HOOZ[1]
Checked by K.-K. CHAN, A. SPECIAN, and A. BROSSI

1. Procedure

A dry, 300-ml., three-necked flask is equipped with a magnetic stirring bar and reflux condenser (to which is attached a Drierite-filled drying tube) and charged with 90 ml. of carbon tetrachloride (Note 1) and 15.42 g. of geraniol (0.10 mole) (Note 2). To this solution is added 34.09 g. of triphenylphosphine (0.13 mole) (Note 3), and the reaction mixture is stirred and heated to reflux for 1 hour. This mixture is allowed to cool to room temperature; dry pentane is added (100 ml.), and stirring is continued for an additional 5 minutes.

The precipitate of triphenylphosphine oxide is filtered and washed with 50 ml. of pentane. The solvent is removed from the combined filtrate at the rotary evaporator under water aspirator pressure at room temperature. Distillation of the residue through a 2-cm. Vigreux column attached to a short-path distillation apparatus (Note 4) provides 13.0–14.0 g. (75–81%) of geranyl chloride, b.p. 47–49° (0.4 mm.), $n^{23}D = 1.4794$ (Note 5).

2. Notes

1. Carbon tetrachloride was dried over magnesium sulfate and distilled from phosphorus pentoxide through a 25-cm. Vigreux column. Lower yields were obtained when either the glassware or reagents were not dried.

2. The geraniol was purchased from Koch-Light Laboratories (>98% pure), dried over potassium carbonate and distilled through

an 8-cm. Vigreux column, b.p. 108–109° (8 mm.). The checkers used geraniol purchased from Aldrich Chemical Co., Inc. and distilled.

3. Triphenylphosphine, m.p. 80–81°, was obtained from Eastman Organic Chemicals and was kept in a drying pistol held at approximately 65° (1 mm.) for approximately 18 hours prior to use. When only 10–20% excess triphenylphosphine was employed the yield of geranyl chloride was approximately 75%, but the small amount of unreacted geraniol which remained rendered product isolation more difficult.

4. A "Bantam-ware" distilling head and condenser assembly from Kontes Glass Co. (K-287100) was used. Foaming may occur due to incomplete removal of solvent. This can be avoided by cooling the distillation flask to approximately −50° and gradually lowering the pressure to 10 mm. The pot temperature is then allowed to increase gradually to room temperature, and the distillation then proceeds without difficulty.

5. The pure geranyl chloride has characteristic infrared absorption (liquid film) at 845, 1255, and 1665 cm.$^{-1}$. The use of these absorptions to assay mixtures of geranyl and linalyl chloride has been discussed in detail.[7] The proton magnetic resonance spectrum (100 MHz., carbon tetrachloride solution) shows absorption at δ 1.61 and 1.67 (2 × 3 singlets, C=C(CH_3)$_2$, 1.71 [3, doublet, J = 1.4 Hz, C=C(CH_3)CH$_2$], 2.05 (4, multiplet, CH_2CH_2), 3.98 (2, doublet, J = 8 Hz, CH_2Cl), 5.02 [1, multiplet, CH=C(CH$_3$)$_2$] and 5.39 p.p.m. (1, triplet of partially resolved multiplets, C=CHCH$_2$Cl.

3. Discussion

Geranyl chloride has been prepared by allylic rearrangement of linalool using hydrogen chloride in toluene solution at 100° or phosphorus trichloride in the presence of potassium carbonate at 0°.[2] The conversion of geraniol to geranyl chloride has been reported using hydrogen chloride in toluene,[2] phosphorus trichloride or phosphorus pentachloride in petroleum ether,[3] and thionyl chloride and pyridine.[4-6] These methods give difficultly separable mixtures[7,8] of geranyl and linalyl chloride, and tedious fractionation[5] is required to isolate the geranyl chloride. Procedures which give a pure product involve treatment of geraniol (a) in ether–hexamethylphosphoramide (HMPA) with methyllithium, followed by p-toluenesulfonyl chloride and lithium chloride in ether–HMPA;[9] (b) in s-collidine with lithium chloride in dimethylformamide, followed by methanesulfonyl chloride;[10] and (c) in pentane with methanesulfonyl chloride at $-5°$, followed by the addition of pyridine.[8]

The present procedure is representative of a fairly general method of converting alcohols to chlorides using carbon tetrachloride and a tertiary phosphine. The reaction occurs under mild, essentially neutral conditions and, as illustrated by the present synthesis, may be employed to convert allylic alcohols to the corresponding halide without allylic rearrangement.

Carbon tetrachloride serves as both solvent and halogen source. Several trivalent phosphorus reagents may be employed, including triphenylphosphine,[11,12] tri-n-octylphosphine,[12] tri-n-butylphosphine,[13] and trisdimethylaminophosphine.[13] The latter "more nucleophilic" phosphines react more rapidly and under milder conditions than triphenylphosphine. When triphenylphosphine is employed, the by-product, triphenylphosphine oxide, usually precipitates completely and is easily removed by filtration. After evaporation of the solvent, the product is isolated in high purity by distillation. On occasion, difficulties may be encountered in separating the alkyl halide from the accompanying oxide. The presence of residual soluble organophosphine oxide may pose a serious problem when attempting to isolate sensitive (*e.g.*, allylic or optically active) halides and can lead to loss of product, racemization, etc. This difficulty is usually resolved, as illustrated in the present synthesis, by adding a diluent such as pentane to ensure precipitation. Although precise conditions will undoubtedly depend on the specific substrate at hand, it is usually desirable to employ a modest excess of the organophosphine. This is especially helpful for the preparation of sensitive halides since, by ensuring complete consumption of alcohol, it simplifies the isolation procedure (Note 3) by avoiding the possible necessity of a careful fractionation step.

Triphenylphosphine reacts with carbon tetrachloride[14] or carbon tetrabromide[15] *in the absence of alcohols* to form the corresponding triphenylphosphine dihalomethylene ylide and triphenylphosphine dihalide.

$$2(C_6H_5)_3P + CX_4 \rightarrow (C_6H_5)_3P\!=\!CX_2 + (C_6H_5)_3PX_2$$
$$X = Cl, Br$$

The mechanism has been viewed as involving either the formation of a pentacovalent phosphorus intermediate,[14,15] or alternatively, by initial nucleophilic attack on halogen

$$(C_6H_5)_3P + CX_4 \rightarrow (C_6H_5)_3P\begin{smallmatrix}X\\CX_3\end{smallmatrix}$$

$$(C_6H_5)_3P + (C_6H_5)_3P\begin{smallmatrix}X\\CX_3\end{smallmatrix} \rightarrow (C_6H_5)_3P{=}CX_2 + (C_6H_5)_3PX_2$$

to form intermediate phosphonium species.[16] These mixtures react with carbonyl compounds to provide a useful route to

$$(C_6H_5)_3P + CX_4 \rightarrow (C_6H_5)_3\overset{+}{P}X\overset{-}{C}X_3$$

$$(C_6H_5)_3\overset{+}{P}X\overset{-}{C}X_3 \rightarrow (C_6H_5)_3\overset{+}{P}CX_3\overset{-}{X}$$

$$(C_6H_5)_3P + (C_6H_5)_3\overset{+}{P}CX_3\overset{-}{X} \rightarrow (C_6H_5)_3P{=}CX_2 + (C_6H_5)_3PX_2$$

1,1-dihaloalkenes.[14,15,17]

The success of the present method depends critically on the *initial* presence of an alcohol to trap the intermediate phosphonium species.[12] If the alcohol is added last, the R_3P—CX_4 reaction described above (an exothermic process for the more nucleophilic phosphines) may go to completion, in which case little or no alkyl halide is formed.[13] Since the reaction displays several characteristics of an S_N2 process, it is thought to proceed by the pathway illustrated:

$$R_3P + CX_4 \rightarrow R_3\overset{+}{P}X\overset{-}{C}X_3$$

$$R_3\overset{+}{P}X\overset{-}{C}X_3 + R'OH \rightarrow R_3\overset{+}{P}O\overset{-}{R'} + HCX_3$$

$$R_3\overset{+}{P}OR' \rightarrow R_3\overset{+}{P}OR'\overset{-}{X}$$

$$R_3\overset{+}{P}OR'\overset{-}{X} \rightarrow R'X + R_3PO$$

Yields of chlorides are good to excellent for primary and secondary alcohols, but a competing olefin-forming elimination process renders the method of limited value for preparing tertiary chlorides.[12] An adaptation of the procedure using carbon tetrabromide allows the synthesis of alkyl bromides. Some examples are the preparation of $n\text{-}C_5H_{11}Br$ (97%) and $C_6H_5CH_2Br$ (96%).[12] Farnesyl bromide has been prepared in 90% yield from farnesol.[23]

The advantages of the carbon tetrahalide–organophosphine–alcohol reaction to prepare halides are simplicity of experimental procedure; good yields; relatively mild, essentially neutral reaction conditions; absence of allylic rearrangements. The reaction proceeds with inversion of configuration and is a useful simple device for converting optically active alcohols to chiral halides in high optical purity.[12,22]

1. Department of Chemistry, University of Alberta, Edmonton, Alberta, Canada, T6G 2E1.
2. J. L. Simonsen and L. N. Owen, "The Terpenes," Vol. I, 2nd ed., Cambridge University Press, 1953, pp. 50, 63.
3. L. Ruzicka, *Helv. Chim. Acta*, **6**, 492 (1923).
4. M. O. Forster and D. Cardwell, *J. Chem. Soc.*, 1338 (1913).
5. D. Barnard and L. Bateman, *J. Chem. Soc.*, 926 (1950).
6. R. M. Carman and N. Dennis, *Aust. J. Chem.*, **20**, 783 (1967).
7. D. Barnard, L. Bateman, A. J. Harding, H. P. Koch, N. Sheppard, and G. B. B. M. Sutherland, *J. Chem. Soc.*, 915 (1950).
8. C. A. Bunton, D. L. Hachey, and J.-P. Leresche, *J. Org. Chem.*, **37**, 4036 (1972).
9. G. Stork, P. A. Grieco, and M. Gregson, *Tetrahedron Lett.*, 1393 (1969); *Org. Syn.*, **54**, 68 (1974).
10. E. W. Collington and A. I. Meyers, *J. Org. Chem.*, **36**, 3044 (1971).
11. I. M. Downie, J. B. Holmes, and J. B. Lee, *Chem. Ind.* (London), 900 (1966); J. B. Lee and T. J. Nolan, *Can. J. Chem.*, **44**, 1331 (1966).
12. J. Hooz and S. S. H. Gilani, *Can. J. Chem.*, **46**, 86 (1968).
13. I. M. Downie, J. B. Lee, and M. F. S. Matough, *Chem. Commun.*, 1350 (1968).
14. R. Rabinowitz and R. Marcus, *J. Amer. Chem. Soc.*, **84**, 1312 (1962).
15. F. Ramirez, N. B. Desai, and N. McKelvie, *J. Amer. Chem. Soc.*, **84**, 1745 (1962).
16. B. Miller, in M. Grayson and E. J. Griffith, eds. "Topics in Phosphorus Chemistry," Vol. 2, Wiley-Interscience, New York, 1965, p. 133.
17. Reagents of the type R_3PX_2 ($R=C_6H_5$, n-C_4H_9, OC_6H_5) have independently been employed to convert alcohols to halides and phenols to aryl halides.[18-20] The reaction of $(C_6H_5O)_3P$ with methyl iodide and an alcohol gives good yields of iodides. Replacing the methyl iodide by benzyl bromide or chloride permits the synthesis of alkyl bromides and chlorides.[21]
18. L. Horner, H. Oediger, and H. Hoffmann, *Justus Liebigs Ann. Chem.*, **626**, 26 (1959).
19. H. Hoffmann, L. Horner, H. G. Wippel, and D. Michael, *Chem. Ber.*, **95**, 523 (1962).
20. G. A. Wiley, R. L. Hershkowitz, B. M. Rein, and B. C. Chung, *J. Amer. Chem. Soc.*, **86**, 964 (1964).
21. H. N. Rydon, *Org. Syn.*, **51**, 44 (1971).
22. D. Brett, I. M. Downie, J. B. Lee, and M. F. S. Matough, *Chem. Ind.* (London), 1017 (1969).
23. E. H. Axelrod, G. M. Milne, and E. E. van Tamelen, *J. Amer. Chem. Soc.*, **92**, 2139 (1970).

ALLYLIC CHLORIDES FROM ALLYLIC ALCOHOLS: GERANYL CHLORIDE

[(E)-1-Chloro-3,7-dimethyl-2,6-octadiene]

geraniol $\xrightarrow{\text{CH}_3\text{Li, CH}_3\text{C}_6\text{H}_4\text{SO}_2\text{Cl}}_{\text{LiCl, HMPA-(C}_2\text{H}_5)_2\text{O, 0°}}$ geranyl chloride

Submitted by GILBERT STORK,[1] PAUL A. GRIECO,[2] and MICHAEL GREGSON
Checked by P. A. ARISTOFF and R. E. IRELAND

1. Procedure

A dry, 1-l., three-necked, round-bottomed flask is equipped with an overhead mechanical stirrer, a 125-ml. pressure equalizing dropping funnel fitted with a rubber septum, and a nitrogen inlet tube. The system is flushed with nitrogen, and 15.4 g. (0.1 mole) of geraniol (Note 1), 35 ml. of dry hexamethylphosphoramide (Note 2), 100 ml. of anhydrous ether (Note 3), and 50 mg. of triphenylmethane (Note 4) are placed in the flask. The stirred solution is cooled to 0° with an ice bath, and 63 ml. (0.1 mole) of 1.6M methyllithium in ether (Note 5) is injected into the addition funnel. The methyllithium solution is added dropwise over a period of 30 minutes. After the addition is complete, the funnel is rinsed by injecting 5 ml. of dry ether.

A solution of 20.0 g. (0.105 mole) (Note 6) of p-toluenesulfonyl chloride in 100 ml. of anhydrous ether is then injected into the addition funnel, and this solution is added over a period of 30 minutes to the red reaction mixture at 0° with stirring. The red color immediately disappears upon addition. After addition is complete, 4.2 g. (0.1 mole) of anhydrous lithium chloride (Note 7) is added. The reaction mixture is warmed to room temperature and stirred overnight (18–20 hours), during which time lithium p-toluenesulfonate precipitates.

After a total of 20–22 hours, 100 ml. of ether is added, followed by 100 ml. of water. The layers are separated, the organic phase is washed four times with 100-ml. portions of water, and finally with 100 ml. of a saturated sodium chloride solution. After drying the organic phase over anhydrous magnesium sulfate, the solvent is removed under reduced pressure on a rotary evaporator. The crude product is transferred to a 50-ml. flask, and the product is distilled through a 20-cm. Vigreux column to give 14.1–14.6 g. (82–85%) of geranyl chloride as a colorless liquid, b.p. 78–79° (3.0 mm.) (Notes 8 and 9).

2. Notes

1. Geraniol, (+99°), can be purchased from the Aldrich Chemical Company, Inc.

2. Hexamethylphosphoramide was purchased from the Fisher Scientific Company and the Aldrich Chemical Company, Inc. It was distilled from calcium hydride prior to use.

3. Anhydrous ether, available from J. T. Baker Chemical Company, can be used without further drying.

4. Available from Eastman Organic Chemicals. Although not necessary, triphenylmethane was used as an indicator to check the molarity of the methyllithium used.

5. Methyllithium (prepared from methyl chloride), available from Foote Mineral Company, can be used without further purification. Attention should be drawn to the following: methyllithium purchased from Alfa Inorganics is prepared from methyl bromide and thus produces a mixture of geranyl bromide and chloride.

6. p-Toluenesulfonyl chloride available from either the Aldrich Chemical Company, Inc. or Matheson, Coleman and Bell, Inc. was used without further purification.

7. Available from Alfa Inorganics. If necessary, finely powdered lithium chloride can be dried by heating under vacuum (0.1 mm.) at 100° for several hours.

8. Our sample of geranyl chloride was identical (i.r., n.m.r., and mass spectrum) with a sample prepared by an alternate route (Professor John Hooz, Department of Chemistry, University of Alberta).

9. The infrared spectrum (neat) shows major absorptions at 2970, 2920, 2855, 1660, 1450, 1375, 1380, 1255, 835, and 660 cm.$^{-1}$ The proton magnetic resonance spectrum (carbon tetrachloride solution, tetramethylsilane reference) has a four-line multiplet in the 1.55–1.85 p.p.m. region characteristic of the olefinic methyl protons, two peaks in the 2.0–2.2 p.p.m. region due to the four allylic methylene protons, a doublet at 4.02 p.p.m. ($J = 7.0$ Hz.) due to the allylic methylene protons adjacent to the chlorine, a very broad triplet at 5.09 p.p.m.,

and a broad triplet at 5.45 p.p.m. ($J = 7.0$ Hz.), both due to the vinyl protons.

3. Discussion

The reaction described here illustrates a general procedure for the preparation of allylic chlorides from allylic alcohols without rearrangement and under conditions allowing the retention of sensitive groups.[3] For example, the sensitive acetal alcohol I with geraniol geometry was similarly treated with ether-hexamethylphosphoramide, with methyllithium in ether, and then with p-toluenesulfonyl chloride and lithium chloride. Workup afforded the corresponding chloride II in 80% yield with no detectable rearrangement. The method was equally successful with the *cis*-isomer of I.

$$(MeO)_2HC\diagup\underset{I}{\overset{CH_3}{\diagdown}}\diagup OH \longrightarrow (MeO)_2HC\diagup\underset{II}{\overset{CH_3}{\diagdown}}\diagup Cl$$

In addition, 85–90% yields of neryl chloride could be obtained from nerol, the geometrical isomer of geraniol. A modification[4] of the above method has appeared which employs methanesulfonyl chloride and a mixture of lithium chloride, dimethylformamide, and collidine at 0°; however, its applicability to compounds possessing sensitive groups was not demonstrated.

Initial attempts at preparing γ,γ-disubstituted allyl chlorides employing thionyl chloride in the presence of tributylamine[5] led to appreciable amounts of rearranged (tertiary) halides.

1. Department of Chemistry, Columbia University, New York, New York 10027.
2. Department of Chemistry, University of Pittsburgh, Pittsburgh, Pennsylvania 15213.
3. G. Stork, P. A. Grieco, and M. Gregson, *Tetrahedron Lett.*, 1393 (1969).
4. E. W. Collington and A. I. Meyers, *J. Org. Chem.*, **36**, 3044 (1971).
5. See W. G. Young, F. F. Caserio, Jr., and D. D. Brandon, Jr., *J. Amer. Chem. Soc.*, **82**, 6163 (1960).

5β-CHOLEST-3-ENE-5-ACETALDEHYDE

$$\text{4-cholesten-3β-ol} + CH_2=CH-OC_2H_5 \xrightarrow[\text{reflux, 17 hrs.}]{Hg(OOCCH_3)_2}$$

$$\text{vinyl ether intermediate} \xrightarrow[\text{5 hrs.}]{220-225°} \text{5β-cholest-3-ene-5-acetaldehyde}$$

Submitted by R. E. IRELAND[1] and D. J. DAWSON
Checked by W. PAWLAK and G. BÜCHI

1. Procedure

A 50-ml., round-bottomed flask equipped with a magnetic stirring bar and a 20-ml. calibration mark (Note 1) is charged with 970 mg. (2.5 mmoles) of 4-cholesten-3β-ol (Note 2). Ethyl vinyl ether is distilled into the flask to the 20-ml. mark (Note 3). The mixture is stirred to effect solution; then 820 mg. (2.55 mmoles) of mercuric acetate (Note 4) is added to the reaction mixture. The flask is fitted with a reflux condenser connected to a gas inlet tube and flushed with argon. The reaction mixture is then stirred and heated (Note 5) at reflux under a positive argon pressure for 17 hours. After the solution has cooled to room temperature, 0.062 ml. (1.09 mmoles) of glacial acetic acid (Note 6) is added, and stirring is continued for 3 hours. The reaction mixture is poured into a preshaken mixture of 150 ml. of petroleum ether (Note 7) and 50 ml. of a 5% aqueous potassium hydroxide solution. The aqueous phase is extracted with 50 ml. of petroleum ether, and the combined extracts are washed with three 50-ml. portions of a 20% aqueous sodium chloride solution and dried over anhydrous sodium carbonate. After the drying agent is removed by filtration, the filtrate is evaporated at reduced pressure (Note 8) to give 1.11 g. of an oil which, upon filtration through 5 g. of silica gel (Note 9) with 200 ml. of petroleum ether, affords 0.81 g. of the cholestenyl vinyl ether as a clear, colorless oil. If desired, crystallization of this oil from 10 ml. of acetone will give 0.74 g. (71%) of the vinyl ether as colorless prisms, m.p. 55–56.5° (Note 10).

Alternatively, the crude vinyl ether (0.81 g.) is transferred with petroleum ether into a 50-ml., round-bottomed flask fitted with a long gas inlet tube. After the petroleum ether is removed at reduced pressure (Note 8), the flask is filled with argon and heated (Note 11) under a positive argon pressure at 220–225° for 5 hours; little or no bubbling should occur. After cooling, the oil is chromatographed on 75 g. of silica gel using 10% ether in petroleum ether as the elution solvent (Notes 7, 9, 12). The first 175 ml. of eluant contains side products and is discarded; elution with another 175 ml. of the solvent gives 0.45–0.55 g. (50–53% overall yield from 4-cholesten-3β-ol) of 5β-cholest-3-ene-5-acetaldehyde as white prisms, m.p. 66.5–68° (Note 10).

2. Notes

1. This flask must be cleaned with hot chromic acid solution and then, along with *all* other glassware used in this preparation, soaked in a base solution, rinsed with distilled water, and oven dried. Thermal rearrangement of the intermediate vinyl ether in a new (untreated) flask resulted in elimination.

2. 4-Cholesten-3β-ol can be prepared by the procedure of Burgstahler and Nordin.[2] A melting point below 130° indicates that the material is contaminated with some of the 3α-hydroxy isomer. The material used above melted at 130.5–131° (from ethanol).

3. Eastman practical grade ethyl vinyl ether was dried over anhydrous sodium carbonate, distilled (b.p. 36°) from sodium wire, and then redistilled from calcium hydride (b.p. 36°) into the reaction flask after a 5-ml. forerun is discarded.

4. Matheson, Coleman, and Bell mercuric acetate was partially dissolved in hot absolute ethanol containing 0.02% glacial acetic acid (Note 6) and filtered by suction. The filtrate was cooled, and the white plates of mercuric acetate were collected by suction filtration and stored under vacuum.

5. An oil bath at 50–55° was found to be satisfactory.

6. DuPont 99.7% acetic acid was used without purification.

7. Baker petroleum ether (b.p. 30–60°) was used.

8. The solvent was removed by rotary evaporation followed by vacuum (0.01 mm.) drying for 1 hour.

9. Merck silica gel (0.05–0.2 mm., 70–325 mesh ASTM) was used. The filtration column (1.4 × 7 cm.) is prepared in the same way as one used for chromatography, only one (200-ml.) fraction is collected. Use of alumina for the filtration gives variable results.

10. Burgstahler and Nordin report the melting point for the vinyl ether as 56–57° and for the aldehyde, 66–90°.[2]

11. A Kügelrohr oven was used.

12. Mallinckrodt anhydrous diethyl ether was used. The chromatography column was 2.7 × 27 cm.

3. Discussion

The Claisen rearrangement[3] has been adapted in recent years to provide a viable synthetic sequence for the preparation of functional groups other than aldehydes and ketones. Ester[4] and amide[5] syntheses have been reported which proceed through the Claisen intermediate (**A**). The Claisen rearrangement has also been used to generate *trans*-trisubstituted double bonds stereoselectively,[4,6–9] angularly-functionalized derivatives,[10] substituted cyclohexenes,[11] acids,[12] and furans.[7]

R = —H, —Alkyl, —Aryl, —OR′, —NR′$_2$, —OSiR′$_3$

A

The procedure given above is an excellent example of the utilization of the Claisen rearrangement to generate an angularly functionalized steroid. The vinyl ether and aldehyde were originally prepared by Burgstahler and Nordin.[2] This procedure combines variations employed by Ireland and co-workers and, in addition, introduces the use of silica gel for the purification of the vinyl ether, thereby improving the reproducibility of the procedure.

1. Division of Chemistry and Chemical Engineering, Gates and Crellin Laboratories of Chemistry, California Institute of Technology, Pasadena, California 91109.

2. A. W. Burgstahler and I. C. Nordin, *J. Amer. Chem. Soc.*, **83**, 198 (1961).
3. P. de Mayo, "Molecular Rearrangements," Part One, Wiley, New York, 1963, pp. 660–684.
4. W. S. Johnson, L. Werthemann, W. R. Bartlett, T. J. Brocksom, T.-t. Li, D. J. Faulkner, and M. R. Peterson, *J. Amer. Chem. Soc.*, **92**, 741 (1970).
5. A. E. Wick, D. Felix, K. Steen, and A. Eschenmoser, *Helv. Chim. Acta*, **47**, 2425 (1964); D. Felix, K. Gschwend-Steen, A. E. Wick, and A. Eschenmoser, *Helv. Chim. Acta*, **52**, 1030 (1969).
6. C. L. Perrin and D. J. Faulkner, *Tetrahedron Lett.*, 2783 (1969).
7. D. J. Faulkner and M. R. Petersen, *J. Amer. Chem. Soc.*, **95**, 553 (1973).
8. R. Marbet and G. Saucy, *Helv. Chim. Acta*, **50**, 2095 (1967).
9. R. I. Trust and R. E. Ireland, *Org. Syn.*, **53**, 116 (1973).
10. R. F. Church, R. E. Ireland, and J. A. Marshall, *J. Org. Chem.*, **31**, 2526 (1966).
11. G. Büchi and J. E. Powell, Jr., *J. Amer. Chem. Soc.*, **92**, 3126 (1970).
12. R. E. Ireland and R. H. Mueller, *J. Amer. Chem. Soc.*, **94**, 5897 (1972).

ETHYL 5β-CHOLEST-3-ENE-5-ACETATE

$$\text{cholesterol} + CH_3C(OC_2H_5)_3 \xrightarrow{\text{heat}} \text{ethyl 5}\beta\text{-cholest-3-ene-5-acetate}$$

Submitted by R. E. Ireland[1] and D. J. Dawson
Checked by W. Pawlak and G. Büchi

1. Procedure

A 100-ml. Claisen distillation flask with two 14/20 standard taper joints and a thermometer inlet is equipped with a gas inlet adapter, a receiver, a thermometer, and a magnetic stirring bar. A 40-ml. calibration mark is made on the flask, and 970 mg. (2.5 mmoles) of cholest-4-en-3β-ol (Note 1) is introduced. Triethyl orthoacetate is then distilled under argon into the flask to the 40-ml. mark (Note 2). The mixture is stirred to effect solution while the flask is purged with argon, and then the top joint is sealed with a thermometer (Figure 1). The stirred solution is heated under a positive pressure of argon so that the vapor reflux level is just below the side arm of the flask; the temperature on the lower thermometer is 142–147°; the upper thermometer temperature is kept between 25 and 70° (Note 3). After 8 days of reflux, during which time a small amount of the volatile material distills into the receiver, the reaction flask is cooled, and all the volatile materials are removed at reduced pressure (Note 4). The residue (1.3 g. of a pale yellow oil) is chromatographed on 120 g. of silica gel with 10% ether in petroleum ether as the eluant (Note 5). The side products eluted with the first 240 ml. of the solvent are discarded; further elution with 120 ml. of the solvent affords 690 mg. of ethyl 5β-cholest-3-ene-5-acetate as a clear, colorless oil. Trituration of this product with acetone produces 560–690 mg. (49–60%) of the ester as white plates, m.p. 89–92.5°.

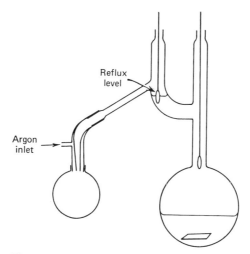

Figure 1.

2. Notes

1. Cholest-4-en-3β-ol can be prepared by the procedure of Burgstahler and Nordin.[2] A melting point below 130° indicates that the material is

contaminated with some of the 3α-hydroxy isomer. The material used above melted at 130.5–131° (from ethanol).

2. The Matheson, Coleman and Bell product was used without purification. After a 10-ml. forerun, the triethyl orthoacetate was distilled (b.p. 142–147°) directly into the reaction flask.

3. A sand bath in an electric heating mantle was found to be satisfactory for the long-term heating process.

4. The volatile materials were removed by rotary evaporation followed by vacuum (0.1 mm.) drying for 1 hour.

5. Merck silica gel (0.05–0.2 mm., 70–325 mesh ASTM) was used in a 3.5 × 26 cm. column. Mallinckrodt anhydrous diethyl ether and Baker petroleum ether (b.p. 30–60°) were employed as eluants.

3. Discussion

The ester–Claisen rearrangement procedure of Johnson and co-workers[3] was modified for use with cholest-4-en-3β-ol.

1. Division of Chemistry and Chemical Engineering, Gates and Crellin Laboratories of Chemistry, California Institute of Technology, Pasadena, California 91109.
2. A. W. Burgstahler and I. C. Nordin, *J. Amer. Chem. Soc.*, **83**, 198 (1961).
3. W. S. Johnson, L. Werthemann, W. R. Bartlett, T. J. Brocksom. T.-t. Li, D. J. Faulkner, and M. R. Peterson, *J. Amer. Chem. Soc.*, **92**, 741 (1970).

N,N-DIMETHYL-5β-CHOLEST-3-ENE-5-ACETAMIDE

[Reaction scheme: cholest-4-en-3β-ol + CH₃—C(OCH₃)₂—N(CH₃)₂ →(o-xylene, heat) N,N-dimethyl-5β-cholest-3-ene-5-acetamide]

Submitted by R. E. IRELAND[1] and D. J. DAWSON
Checked by W. PAWLAK and G. BÜCHI

1. Procedure

A 50-ml., round-bottomed flask, equipped with a Teflon®-covered magnetic stirring bar and a reflux condenser connected to a gas inlet tube, is charged with 970 mg. (2.5 mmoles) of cholest-4-en-3β-ol (Note 1) and 30 ml. of o-xylene (Note 2). The mixture is stirred to effect solution, and then 1.67 g. (0.0125 mole) of N,N-dimethylacetamide dimethylacetal (Note 3) is added. The flask is flushed with argon and then heated (Note 4) at reflux under a positive pressure of argon with vigorous stirring for 65 hours. After cooling, the volatile materials are removed at reduced pressure (Note 5), and the yellow oily residue (1.2 g.) is chromatographed on 60 g. of silica gel with ether (Note 6). Elution of the column with 200 ml. of ether gives a mixture of cholestadienes which is discarded; further elution with 500 ml. of ether affords 740 mg. of N,N-dimethyl-5β-cholest-3-ene-5-acetamide as a clear, colorless oil which on trituration with acetone gives 740 mg. (65%) of the amide as white plates, m.p. 128–129.5°.

2. Notes

1. Cholest-4-en-3β-ol can be prepared by the procedure of Burgstahler and Nordin[2]. A melting point below 130° indicates that the material is contaminated with some of the 3α-hydroxy isomer. The material used above melted at 130.5–131° (from ethanol).
2. The Matheson, Coleman and Bell product was used without purification.
3. N,N-Dimethylacetamide dimethylacetal was obtained from Fluka A. G. and used without purification.
4. A sand bath set into an electric heating mantle was found to be satisfactory for the long-term heating process.
5. The volatile materials were removed by rotary evaporation followed by vacuum (0.1 mm.) drying for 1 hour.
6. Merck silica gel (0.05–0.2 mm., 70–325 mesh ASTM) was used in a 2.5 × 25 cm. column. Mallinckrodt anhydrous diethyl ether was employed as the eluant.

3. Discussion

The amide–Claisen rearrangement procedure of Eschenmoser and co-workers[3] was modified for use with cholest-4-en-3β-ol.

1. Division of Chemistry and Chemical Engineering, Gates and Crellin Laboratories of Chemistry, California Institute of Technology, Pasadena, California 91109.
2. A. W. Burgstahler and I. C. Nordin, *J. Amer. Chem. Soc.*, **83,** 198 (1961).
3. A. E. Wick, D. Felix, K. Steen, and A. Eschenmoser, *Helv. Chim. Acta*, **47,** 2425 (1964).

PREPARATION OF VINYL TRIFLUOROMETHANESULFONATES: 3-METHYL-2-BUTEN-2-YL TRIFLATE

(Methanesulfonic acid, trifluoro-, 1,2-dimethylpropenyl ester)

$$CF_3SO_3H \xrightarrow[-H_2O]{P_2O_5} (CF_3SO_2)_2O$$

$$(CH_3)_2CHCOCH_3 + (CF_3SO_2)_2O + \underset{N}{\bigcirc} \xrightarrow[-78 \text{ to } 25°]{\text{pentane}}$$

$$(CH_3)_2C=\underset{OSO_2CF_3}{\overset{|}{C}CH_3} + (CH_3)_2CH\underset{OSO_2CF_3}{\overset{|}{C}=CH_2} + \underset{}{\bigcirc}NH^{\oplus} \quad {}^{\ominus}OSO_2CF_3$$

Submitted by PETER J. STANG[1] and THOMAS E. DUEBER
Checked by WAYNE JAEGER and HERBERT O. HOUSE

1. Procedure

A. *Trifluoromethanesulfonic Anhydride.* To a dry, 100-ml., round-bottomed flask are added 36.3 g. (0.242 mole) of trifluoromethanesulfonic acid (Note 1) and 27.3 g. (0.192 mole) of phosphorus pentoxide (Note 2). The flask is stoppered and allowed to stand at room temperature for at least 3 hours. During this period the reaction changes from a slurry to a solid mass. The flask is fitted with a short-path distilling head and then heated first with a stream of hot air from a heat gun and then with the flame from a small burner. The flask is heated until no more trifluoromethanesulfonic anhydride distills, b.p. 82–115°. The yield of the anhydride, a colorless liquid, is 28.4–31.2 g. (83–91%). Although this product is sufficiently pure for use in the next step of this preparation, the remaining acid may be removed from the anhydride by the following procedure. A slurry of 3.2 g. of phosphorous pentoxide in 31.2 g. of the crude anhydride is stirred at room temperature in a stoppered flask for 18 hours. After the reaction flask has been fitted with a short-path distilling head, it is heated with an oil bath to distill 0.7 g. of forerun, b.p. 74–81°, followed by 27.9 g. of the pure trifluoromethanesulfonic acid anhydride, b.p. 81–84° (Note 3).

B. *3-Methyl-2-buten-2-yl Triflate.* A solution of 2.58 g. (0.030

mole) of 3-methyl-2-butanone (Note 4) and 2.78 g. (0.035 mole) of anhydrous pyridine (Note 5) in 10 ml. of anhydrous pentane (Note 6) is placed in a dry, 50-ml. Erlenmeyer flask, and the flask is stoppered with a rubber septum. After the solution has been cooled in a dry ice–acetone cooling bath, 9.72 g. (5.5 ml., 0.034 mole) of trifluoromethanesulfonic anhydride is added from a hypodermic syringe, dropwise and with swirling during 2–3 minutes. The resulting mixture, from which a white solid separates initially, is allowed to warm to room temperature and stand for 22–24 hours. During this period the reaction mixture becomes red in color (Note 7), and a viscous, red semisolid separates. The supernatant pentane solution is decanted, and the residual viscous semisolid is washed with two 10-ml. portions of pentane. While these combined pentane solutions are stored over anhydrous potassium carbonate, the remaining red semisolid is dissolved in 5 ml. of saturated aqueous sodium bicarbonate and extracted with three 5-ml. portions of pentane. The combined pentane solutions (including the solid potassium carbonate) are washed rapidly (Note 8) with 5 ml. of cold saturated aqueous sodium bicarbonate and then dried over anhydrous potassium carbonate. After the orange pentane solution has been concentrated to a volume of approximately 10 ml. with a rotary evaporator, it is transferred to a small distilling apparatus, and the remainder of the pentane is removed by distillation at atmospheric pressure (Note 9). The residual liquid is fractionally distilled under reduced pressure to separate 2.94–2.97 g. (45%) of a fraction containing mixtures of 3-methyl-2-buten-2-yl triflate and lesser amounts of 3-methyl-1-buten-2-yl triflate as a colorless liquid, b.p. 58–66° (22 mm.), n^{25}D 1.3832–1.3898 (Note 10). Fractional distillation of this mixture with a 40-cm. spinning band column separated higher boiling fractions, b.p. 45–47° (12 mm.), that contained (Note 10) 98% of the 3-methyl-2-buten-2-yl triflate as a colorless liquid, n^{25}D 1.3838 (Note 11).

2. Notes

1. Trifluoromethanesulfonic acid, purchased from Eastman Organic Chemicals, was used without purification. Trifluoromethanesulfonic acid in large quantities is available from 3M Company as Fluorocarbon Acid FC-24.

2. The phosphorus pentoxide should be protected from atmospheric moisture by weighing this reagent in a dry, stoppered flask.

3. This product exhibits one major gas chromatographic peak (retention time 2.3 minutes, silicone fluid QF_1 on Chromosorb P) as well as one minor, unidentified, more rapidly eluted impurity. The product has strong infrared absorption (CCl_4 solution) at 1470, 1240, and 1130 cm.$^{-1}$.

4. 3-Methyl-2-butanone, purchased from Eastman Organic Chemicals, was used without purification. In general, commercially available ketones may be used without further purification.

5. A reagent grade of pyridine, purchased from Fisher Scientific Company, was dried over anhydrous potassium carbonate and distilled. The pyridine was collected at 112–113°.

6. A commercial sample of n-pentane was distilled from calcium hydride to separate the pure solvent, b.p. 35–36°.

7. The intensity of color developed in the reaction mixture is an approximate indication of the extent of reaction. With reactive triflates a dark, almost tarry-looking mass develops in a few hours, while with the slower forming triflates several days at room temperature may be required for adequate color development. In no case was any product isolated when fairly dark color had not developed in the reaction mixture.

8. Only relatively stable vinyl triflates should be washed with water. More reactive triflates such as $CH_2{=}C(C_6H_5)OSO_2CF_3$ do not survive washing with water.

9. For the more volatile triflates, removal of the solvent should be accomplished by distillation to minimize loss of the volatile product. Also, care should be taken not to overheat the residual product since overheating can result in decomposition.

10. The fractions from this distillation may be analyzed by gas chromatography employing a column packed with Carbowax 20M suspended on Chromosorb P. The retention times for the various components (minutes) are: pentane, 1.6; 3-methyl-2-butanone, 4.2; 3-methyl-1-buten-2-yl triflate, 6.7; and 3-methyl-2-buten-2-yl triflate, 9.5.

11. The pure, more highly substituted olefin, $n^{25}{\rm D}$ 1.3840, could also be separated by preparative gas chromatography, and the unchanged ketone could be separated from the two triflate isomers by chromatography on a silica gel column with pentane as the eluant. The pure product has infrared absorption (CCl_4 solution) at 1700 (enol $C{=}C$), 1210 and 1140 cm.$^{-1}$ (SO_2) with end absorption in the ultraviolet (heptane

solution, ϵ 700 at 210 mμ) and the following broad singlets in the proton magnetic resonance spectrum (C_6H_6 solution): δ 1.82 (3H, CH_3), 1.63 (3H, CH_3), and 1.42 (3H, CH_3). The mass spectrum of the product exhibits the following relatively abundant peaks, m/e (relative intensity): 218 (M^+, 58), 69 (75), 57 (92), 43 (100), 41 (44), and 39 (24).

3. Discussion

Vinyl trifluoromethanesulfonates (triflates) are a new class of compounds, unknown before 1969, that have been used most extensively in solvolytic studies to generate vinyl cations.[2,3,8–12] Three methods have been used to prepare these sulfonic esters. The first, involving the preparation and decomposition of acyltriazines,[4] requires several steps to prepare the acyltriazines and is limited to the preparation of fully substituted vinyl triflates. The second method involves the electrophilic addition of trifluoromethanesulfonic acid to acetylenes[5,6,15] and, consequently, is not applicable to the preparation of trisubstituted vinyl triflates and certain cyclic vinyl triflates. However, this second procedure is relatively simple and often gives purer products in higher yield than the subsequently discussed reaction with ketones. Table I lists vinyl triflates that have been prepared by this procedure.

The third procedure illustrated by this preparation involves the reaction of ketones with trifluoromethanesulfonic anhydride in a solvent such as pentane, methylene chloride, or carbon tetrachloride and in the presence of a base such as pyridine, lutidine, or anhydrous sodium carbonate.[7–11,15] This procedure, which presumably involves either acid-catalyzed or base-catalyzed enolization of the ketone followed by acylation of the enol with the acid anhydride, has also been used to prepare other vinyl sulfonate esters such as tosylates[12] or methanesulfonates.[13]

TABLE I
VINYL TRIFLATES PREPARED BY THE REACTION OF ACETYLENES WITH TRIFLUOROMETHANESULFONIC ACID

Substrate	Product	b.p.	Yield, %
$CH_3C\equiv CH$	$CH_2=C(CH_3)OSO_2CF_3$	25–27° (12 mm.)	60–80
$C_2H_5C\equiv CC_2H_5$	cis and trans $C_2H_5CH=C(C_2H_5)OSO_2CF_3$	68.5–69.5° (25 mm.)	40–60
$(CH_3)_3CC\equiv CH$	$CH_2=C(OSO_2CF_3)C(CH_3)_3$	45–50° (15 mm.)	40–60
$C_6H_5C\equiv CH$	$CH_2=C(C_6H_5)OSO_2CF_3$	44–45° (0.3 mm.)	20–60
$CH_3(CH_2)_3C\equiv CH$	$CH_2=C(OSO_2CF_3)CH_2(CH_2)_2CH_3$	67–69° (15 mm.)	70
$(CH_3)_2CHC\equiv CH$	$CH_2=C(OSO_2CF_3)CH(CH_3)_2$	37–40° (15 mm.)	62

TABLE II
Vinyl Triflates Prepared from Ketones

Ketone	Reaction Conditions	Time, Days	Product	Yield, %
$XC_6H_4COCH_3$ (X = H, p-Cl, m-Cl, p-CF_3, p-NO_2)	CH_2Cl_2, Na_2CO_3	3–21	$XC_6H_4\overset{OSO_2CF_3}{\underset{\|}{C}}=CH_2$	15–45
2,6-dimethylcyclohexanone	CH_2Cl_2, Na_2CO_3	14	(2,6-dimethyl-1-cyclohexenyl triflate)	28
	CCl_4, pyridine	5	(isomeric 2,6-dimethyl cyclohexenyl triflate)	54
$C_6H_5CH(CH_3)COCH_3$	CCl_4, pyridine	2–4	$C_6H_5C(CH_3)=C(CH_3)OSO_2CF_3$ (cis, 18%; trans, 51%) + $C_6H_5CH(CH_3)C(OSO_2CF_3)=CH_2$ (31%)	53
	CCl_4, P_2O_5	1	$C_6H_5C(CH_3)=C(CH_3)OSO_2CF_3$ (trans, 80%; cis, 20%)	32
$(CH_3)_2CHCOCH_3$	CH_2Cl_2, Na_2CO_3	10	$(CH_3)_2C=C(CH_3)OSO_2CF_3$ (**1**, 70%) + $(CH_3)_2CHC(OSO_2CF_3)=CH_2$ (**2**, 30%)	33
	CCl_4, pyridine	1	**1** (90%) + **2** (10%)	58

Vinyl tosylates (but not vinyl triflates) may also be prepared from the ditosylate derivatives of 1,2-diols.[14]

Examples of the use of this procedure to prepare vinyl triflates from ketones are provided in Table II. Often mixtures of *cis* and *trans* isomers as well as the various double bond isomers of vinyl triflates are obtained by this procedure, and the amounts of these isomers produced may vary with the base and solvent used. Also, small amounts of unchanged ketone may contaminate the initial crude product. Consequently, separation procedures such as preparative gas chromatography or efficient fractional distillation may be required to obtain a single vinyl triflate isomer.

1. Department of Chemistry, the University of Utah, Salt Lake City, Utah 84112.
2. M. Hanack, *Accounts Chem. Res.*, **3**, 209 (1970).
3. P. J. Stang, *Progr. Phys. Org. Chem.*, **10**, 205 (1973).

4. W. M. Jones and D. D. Maness, *J. Amer. Chem. Soc.*, **91**, 4314 (1969).
5. P. J. Stang and R. H. Summerville, *J. Amer. Chem. Soc.*, **91**, 4600 (1969).
6. A. G. Martinez, M. Hanack, R. H. Summerville, P. v. R. Schleyer, and P. J. Stang, *Angew. Chem., Int. Ed. Engl.*, **9**, 302 (1970).
7. T. E. Dueber, P. J. Stang, W. D. Pfeifer, R. H. Summerville, M. A. Imhoff, P. v. R. Schleyer, K. Hummel, S. Bocher, C. E. Harding, and M. Hanack, *Angew. Chem., Int. Ed. Engl.*, **9**, 521 (1970).
8. W. D. Pfeifer, C. A. Bahn, P. v. R. Schleyer, S. Bocher, C. E. Harding, K. Hummel, M. Hanack, and P. J. Stang, *J. Amer. Chem. Soc.*, **93**, 1513 (1971).
9. M. A. Imhoff, R. H. Summerville, P. v. R. Schleyer, A. G. Martinez, M. Hanack, T. E. Dueber, and P. J. Stang, *J. Amer. Chem. Soc.*, **92**, 3802 (1970).
10. T. C. Clarke, D. R. Kelsey, and R. G. Bergman, *J. Amer. Chem. Soc.*, **94**, 3626 (1972).
11. P. J. Stang and T. E. Dueber, *J. Amer. Chem. Soc.*, **95**, 2683 (1973).
12. N. Frydman, R. Bixon, M. Sprecher, and Y. Mazur, *Chem. Commun.*, 1044 (1969).
13. W. E. Truce and L. K. Liu, *Tetrahedron Lett.*, 517 (1970).
14. P. E. Peterson and J. M. Indelicato, *J. Amer. Chem. Soc.*, **90**, 6515 (1968).
15. R. H. Summerville, C. A. Senkler, P. v. R. Schleyer, T. E. Dueber, and P. J. Stang, *J. Amer. Chem. Soc.*, **96**, 1100 (1974).

CYCLOBUTANONE VIA SOLVOLYTIC CYCLIZATION

$$HC \equiv C-CH_2-CH_2OH + (CF_3SO_2)_2O \xrightarrow{Na_2CO_3} HC \equiv C-CH_2-CH_2OSO_2CF_3$$

$$HC \equiv C-CH_2-CH_2OSO_2CF_3 \xrightarrow[2.\ H_2O]{1.\ CF_3CO_2H + CF_3CO_2Na} \square=O$$

Submitted by M. HANACK,[1] T. DEHESCH, K. HUMMEL, and A. NIERTH
Checked by H. ONA, B. A. BOIRE, and S. MASAMUNE

1. Procedure

A. *3-Butyn-1-yl Trifluoromethanesulfonate.* A 500-ml., three-necked flask is fitted with a mechanical stirrer, a pressure-equalizing dropping funnel, and a stopper. The system is flushed with nitrogen through a gas-inlet tube attached to the top of the funnel. To 150 ml. of dry methylene chloride (Note 1) in the flask is added 75 g. (0.27 mole) of trifluoromethanesulfonic acid anhydride (Note 2), and the solution is cooled to −40°. After addition of 14.5 g. (0.14 mole) of finely powdered anhydrous sodium carbonate (Note 3), 15 g. (0.21 mole) of 3-butyn-1-ol (Note 4) is added dropwise over a 20-minute period to the well-stirred reaction mixture maintained at −40° to −55°. Stirring is continued at −30° for 2 hours and then at 0° for another hour, and finally the reaction is quenched by dropwise addition of 50 ml. of water. The organic layer is separated and dried over anhydrous sodium sulfate, and after filtration the solvent is removed with a rotary evaporator. The temperature of the water bath should not exceed 25°. The resulting residue is placed in a flask directly connected with a liquid nitrogen trap and distilled at 1 Torr. The fractions boiling in the range from 40° to 50° are sufficiently pure for use in the next step. The yield of the sulfonate is 38.94 g. (90%) (Notes 5 and 6).

B. *Cyclobutanone.* A 500-ml., thick-walled ampoule is charged with 210 g. (137 ml.) of trifluoroacetic acid, 11.5 g. (0.085 mole) of sodium trifluoroacetate (Note 7), and 17 g. of 3-butyn-1-yl trifluoromethanesulfonate in this order. A magnetic stirring bar is added, and the ampoule is sealed. The stirred reaction mixture is immersed in a constant temperature bath kept at 65° ($\pm 2°$) for 1 week. The ampoule is cooled slowly to $-50°$ with a dry ice–methanol bath (Note 8) and then opened. With the aid of 200 ml. of ether the reaction mixture is transferred to a 1-l. Erlenmeyer flask to which 74 g. (1.83 mole) of sodium hydroxide in 150 ml. of water is added carefully. During the addition the flask is immersed in the bath maintained at approximately $-50°$ (Note 8). The ethereal layer is separated, and the aqueous layer is saturated with sodium chloride and extracted twice with ether. The original organic layer and ethereal extracts are combined, dried over anhydrous sodium sulfate, and directly distilled into a liquid nitrogen trap. The total condensate in the trap is placed in a distillation flask attached to a 40-cm. Vigreux column and a condenser cooled to $-40°$ by means of a circulating cold bath (Note 9). After the ether is distilled, all volatile materials are collected by raising the bath temperature to 130° to yield 1.84–2.05 g. (31-36%) of cyclobutanone. The purity of the product is more than 95% by nuclear magnetic resonance spectroscopy, the only impurity being diethyl ether (Note 10).

2. Notes

1. The submitters treated methylene chloride with sulfuric acid and then with sodium hydroxide and distilled it before use. The checkers used the reagent grade solvent supplied by Fisher Scientific Company after the overnight storage over Molecular Sieves (4A).

2. The submitters prepared the anhydride, following basically the procedure described by Burdon, Farazmand, Stacey, and Tatlow.[2]

32.1 g. of trifluoromethanesulfonic acid (supplied by Minnesota Mining and Manufacturing Company) maintained at 0°, is added 25 g. of phosphorus pentoxide in three portions, and the resulting anhydride is distilled by gradually heating the reaction mixture to a bath temperature of 110° over a 1-hour period. The fractions boiling at 80–100° (760 torr) are collected and redistilled from approximately 8 g. of phosphorus pentoxide until the distillate no longer fumes on exposure to air. Normally three distillations are necessary. The presence of the trifluoromethanesulfonic acid can be detected by dipping a glass rod into the distillate and waving the wet rod in the air. The anhydride, in contrast, does not fume. The final yield of the product is 25 g. (83%), b.p. 84°. The checkers purchased the anhydride from Pierce Chemical Company and used it without further purification.

3. Anhydrous sodium carbonate was ground into fine powder and dried in vacuum at 200° for 4 hours.

4. In one experiment the checkers used 3-butyn-1-ol available from Aldrich Chemical Company, Inc., and found that it was of satisfactory purity. In other experiments, both the submitters and the checkers prepared the hydroxy compound from sodium acetylide and ethylene oxide in liquid ammonia according to the procedure described by Schulte and Reiss[3] and further attempted to maximize the yield by varying the ratio of sodium:ethylene oxide:liquid ammonia used in the reaction. Unfortunately, the checkers failed to obtain consistent results in repeated experiments and consequently could not define the optimum conditions for the reaction. Thus, the yield of 3-butyn-1-ol varied from 15 to 45% and 15 to 31% on the basis of sodium and ethylene oxide, respectively. Unknown and apparently subtle experimental factors affect the yield significantly.

5. When 3-butyn-1-ol was added to a solution of the anhydride cooled to 0° and then the mixture was allowed to react at room temperature for 3 hours, the yield of the sulfonate dropped to 70–75%.

6. Proton magnetic resonance of 3-butyn-1-ol trifluoromethanesulfonate (carbon tetrachloride) δ, coupling constant J in Hz., number of protons: 2.05 (triplet, $J_{2,4} = 2.6$, 1), 2.76 (doublet of triplets, $J_{1,2} = 6.7$, $J_{2,4} = 2.6$, 2), 4.57 (triplet, $J_{1,2} = 6.7$, 2).

7. The salt is available from Aldrich Chemical Company Inc. However, the checkers readily prepared it in the following way. To a stirred solution of 8.8 g. (0.22 mole) of sodium hydroxide in 400 ml. of 98% ethanol was added dropwise 25 g. (0.22 mole) of trifluoroacetic acid.

After the addition was completed, the ethanol was removed under reduced pressure, and the residue was suspended in approximately 100 ml. of ether, filtered, and washed several times with ether. The yield was 25 g. (83.8%), and the salt was dried to a constant weight in a vacuum desiccator containing calcium sulfate (2 days).

8. Due to the high volatility of cyclobutanone, a substantial amount of the product is lost unless the mixture is sufficiently cooled during the process of neutralization.

9. The cooling to $-40°$ was necessary to prevent the loss of highly volatile cyclobutanone.

10. The checkers redistilled this product through a 3-cm. column to determine its b.p. to be 96.5–97.5° (710 mm.); infrared (chloroform), cm.$^{-1}$: 1780; proton magnetic resonance (carbon tetrachloride) δ, number of protons: 3.05 (multiplet, 4), 1.96 (multiplet, 2).

3. Discussion

Cyclobutanone has been prepared by (1) reaction of diazomethane with ketene,[4] (2) treatment of methylenecyclobutane with performic acid, followed by cleavage of the resulting glycol with lead tetraacetate,[5] (3) ozonolysis of methylenecyclobutane,[6] (4) epoxidation of methylenecyclopropane followed by acid-catalyzed ring expansion,[7] and (5) oxidative cleavage of cyclobutane trimethylene thioketal, which in turn is prepared from 2-(ω-chloropropyl)-1,3-dithiane.[8]

The present procedure[9] represents another synthesis of cyclobutanone through the unique acetylenic bond participation in solvolysis. Cyclobutane derivatives prepared in this way include 2-methyl-, 2-ethyl-, 2-isopropyl-, and 2-trifluoromethylcyclobutanone from the corresponding acetylenic compounds.[10]

1. Institut für Organische Chemie der Universität des Saarlandes, D-66 Saarbrücken, Germany.
2. J. Burdon, I. Farazmand, M. Stacey, and J. C. Tatlow, *J. Chem. Soc.*, 2574 (1957).
3. K. E. Schulte and K. P. Reiss, *Chem. Ber.*, **86**, 777 (1953).
4. P. Lipp and R. Köster, *Chem. Ber.*, **64**, 2823 (1931).
5. J. D. Roberts and C. W. Sauer, *J. Amer. Chem. Soc.*, **71**, 3925 (1949).
6. J. M. Conia, P. Leriverend, and J. L. Ripoll, *Bull. Soc. Chim. Fr.*, 1803 (1961).
7. J. R. Salaün and J. M. Conia, *Chem. Commun.*, 1579 (1971).
8. D. Seebach, N. R. Jones, and E. J. Corey, *J. Org. Chem.*, **33**, 300 (1968); D. Seebach and A. K. Beck, *Org. Syn.*, **51**, 76, (1971).
9. K. Hummel and M. Hanack, *Justus Liebigs Ann. Chem.*, **746**, 211 (1971).
10. M. Hanack, S. Bocher, I. Herterich, K. Hummel, and V. Vött, *Justus Liebigs Ann. Chem.*, **733**, 5 (1970). See also references cited therein.

MACROCYCLIC DIIMINES: 1,10-DIAZACYCLOÖCTADECANE

A. $ClC(CH_2)_6CCl + 2\ H_2N(CH_2)_8NH_2 \xrightarrow[25°]{C_6H_6}$

$$\begin{array}{c} O \quad\quad O \\ \parallel \quad\quad \parallel \\ C(CH_2)_6C \\ NH \quad\quad NH \\ (CH_2)_8 \end{array}$$

$+\ H_2N(CH_2)_8NH\cdot 2\ HCl$

B.

$$\begin{array}{c} O \quad\quad O \\ \parallel \quad\quad \parallel \\ C(CH_2)_6C \\ NH \quad\quad NH \\ (CH_2)_8 \end{array} \xrightarrow[\text{tetrahydrofuran, reflux}]{LiAlH_4} \begin{array}{c} (CH_2)_8 \\ NH \quad\quad NH \\ (CH_2)_8 \end{array}$$

Submitted by CHUNG HO PARK and HOWARD E. SIMMONS[1]
Checked by STEPHEN R. WILSON and ROBERT E. IRELAND

1. Procedure

A. *1,10-Diazacyclooctadecane-2,9-dione.* A 12-l., 4-necked, round-bottomed flask with four indents is fitted with a mechanical stirrer, two dropping funnels (Note 1), and an inlet tube to maintain a static nitrogen atmosphere throughout the reaction. Four liters of benzene (Note 2) are placed in the flask and stirred vigorously, and two solutions containing 33.8 g. (0.16 mole) of suberyl dichloride (Note 3) in 2.0 l. of benzene and 48.2 g. (0.32 mole) of octamethylenediamine (Note 4) in 2.0 l. of benzene are added simultaneously over a 6–7 hour period at room temperature. After the addition is complete, the resulting suspension is stirred slowly overnight. The addition funnels are removed from the flask and replaced by a stopper fitted with a tube of suitable dimensions to permit the reaction mixture to be siphoned from the reaction flask when a slight positive nitrogen pressure is present in the flask. The fine suspension in the reaction flask is agitated and siphoned into a large fritted filter funnel. The white solid is washed three times with benzene and dried in a vacuum oven. The resulting white solid is pulverized and placed in a continuous extractor (Note 5) and extracted for three days with 1 l. of boiling benzene in a 2-l, round-bottomed flask equipped with a magnetic stirring bar and heating mantle. After the extractor is allowed to cool to room temperature, filtration of the white solid suspension in the benzene gives 23.1–23.6 g. (51–52%) of crude 1,10-diazacyclooctadecane-2,9-dione, m.p. 198–201°. The crude product is used for the subsequent step without further purification (Note 6).

B. *1,10-Diazacyclooctadecane.* To a 500-ml., round-bottomed flask equipped with a mechanical stirrer, condenser, and nitrogen bubbler is added 150 ml. of dry tetrahydrofuran (Note 7) and 3.8 g. (0.1 mole; 33% excess) of lithium aluminum hydride. While the suspension is stirred, 14.1 g. (0.05 mole) of 1,10-diazacyclooctadecane-2,9-dione is added in small portions through Gooch tube (Note 8). When the addition is complete and evolution of hydrogen subsides, the mixture is heated at reflux under a nitrogen atmosphere for 48 hours. The mixture is cooled to 5–10° in an ice bath and decomposed by cautious dropwise addition of 3.8 ml. of water followed by 3.8 ml. of 15% sodium hydroxide and finally by 11.8 ml. of water. The mixture is allowed to come to room temperature, stirred for an additional hour, and filtered through a fritted-glass funnel. The resulting cake is washed thoroughly with three 50-ml. portions of tetrahydrofuran followed by three 50-ml. portions of ether. The combined filtrate is concentrated under reduced pressure. After purging with benzene to remove traces of water and then pumping under vacuum overnight, there remains 12.7 g. crude diimine as a colorless, waxy solid.

The crude product at this point contains a small amount of unreduced carbonyl group which can be detected in the infrared spectrum. Treatment of the crude diimine again under the identical reduction conditions (Note 9) using the same amounts of the reagent and solvent gives 10.6 g. (83%) of product. Pure 1,10-diazacyclooctadecane is obtained by distillation of this product through a semimicro Vigreux column (10-cm.) under reduced pressure, b.p. 120–130° (0.05 mm.), 9.5 g. (75%), m.p. 59–61° (Note 10).

2. Notes

1. Two graduated dropping funnels of 500-ml. capacity equipped with pressure-equalizing tubes and small diameter tips to allow two solutions to be added in fine streams were used. They were also equipped with screw-in type plungers with Teflon® fluorocarbon tips to allow fine adjustment of the flow rates. The checkers utilized standard 500-ml. dropping funnels and two needle valve adapters shown in Figure 1.

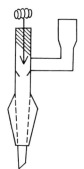

Figure 1.

It is important that the two solutions are added at the same slow rate. The submitters report that mixing of the two solutions in one portion halved the yield of the dilactam.

2. All of the benzene used was dried over sodium wire and distilled from sodium under a nitrogen atmosphere.

3. Suberyl dichloride purchased from Frinton Laboratories was used without further purification. The purity of the acid chloride was checked by gas chromatography by first converting it to the diethyl ester.

4. Octamethylenediamine was purchased from Columbia Organic Chemicals Co., Inc. The purity was checked by gas chromatography and varied from batch to batch, and only samples which were shown to be homogeneous were used. The diamine should be handled in an inert atmosphere, as it forms a carbonate rapidly.

5. An extractor of 2-l. capacity with fritted disc of 85 mm. diameter and coarse porosity was used (Figure 2).

The desired product is only slightly soluble in benzene. Depending

Figure 2.

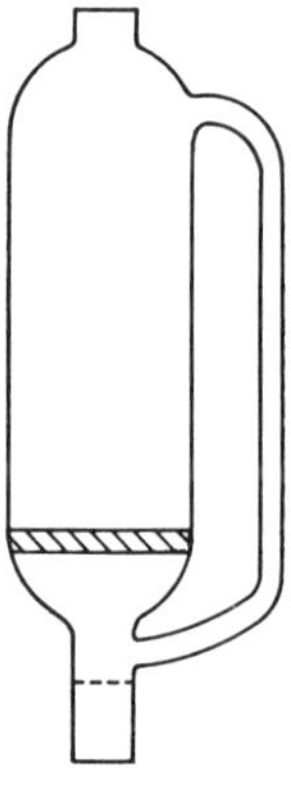

on the efficiency of extraction, this step may be extended to longer periods with beneficial results. Lower yield of the dilactam generally indicates imcomplete extraction. An attempt to isolate the product by washing away the dihydrochloride from the solid mixture with water resulted in the formation of an emulsion.

6. A sample of the crude dilactam dissolved in an ethanol–water mixture gave no precipitate with silver nitrate, indicating that the product is not contaminated with octamethylenediamine dihydrochloride. If the purity is questionable, the crude product can be conveniently purified at this point by recrystallization from dimethylformamide or 50% ethanol in water to give better than 90% recovery of pure product, m.p. 207–209°.

7. Commercially available tetrahydrofuran was distilled from $LiAlH_4$ prior to use.

8. The submitters suggest that on ten times this scale, the reaction flask be purged with a stream of nitrogen during the addition of amide.

9. The repeated treatment with lithium aluminum hydride was necessary to obtain complete reduction, since use of the reducing agent in manyfold excess and longer reaction times failed to accomplish complete reduction in one step. If the product was purified by distillation without the second reduction, the yield of pure diimine amounted to 9.1 g. (71%). The submitters report that on ten times the scale the yield with one reduction was 81 g. (64%).

10. The product obtained is analytically pure, and the submitters claim it is suitable for polymerization. It is also shown to be homogeneous by gas chromatography.

3. Discussion

1,10-Diazacyclooctadecane has also been prepared by another general method which involves reaction of 1,8-dibromooctane and N,N'-ditosyl-1,8-octanediamine with potassium carbonate as base, followed by cleavage of the sulfonamide groups;[2] the yield, however, is not as good as by the present method. The present procedure is essentially a scale-up of the general method used by Stetter and Marx[3] for preparing a series of macrocyclic diimines of 10- to 21-membered rings in 25–78% yield for the crucial cyclization step. The procedure is adaptable for even larger scale preparations using 5 moles of the acid chloride and a 100-gallon reaction kettle.

The reaction is generally applicable to further extension for making macrobicyclic diamines[4] with bridgehead nitrogen atoms by using monocyclic diimines and an appropriate acid chloride. The main

$$HN\underset{(CH_2)_l}{\overset{(CH_2)_k}{\diagdown}}NH + ClC(CH_2)_{m-2}CCl \xrightarrow{Et_3N} N-(CH_2)_l-N\underset{O}{\overset{(CH_2)_k}{\diagdown}}\underset{C-(CH_2)_{m-2}-C}{\overset{||}{||}}$$

$$\xrightarrow{LiAlH_4} N-(CH_2)_l-N\underset{(CH_2)_m}{\overset{(CH_2)_k}{\diagdown}} \quad k, l, m = 6{-}14$$

difference in the procedures is that triethylamine is used as the hydrogen chloride acceptor in the cyclization step instead of using an extra mole of diimine. Triethylamine hydrochloride does not interfere with purification in this case since the bicyclic diamides are fairly soluble in benzene but insoluble in water. The bicyclic diamides are isolated by first removing triethylamine hydrochloride from the reaction mixture by filtration and water washings, followed by evaporation of the solvent. If triethylamine is used as the base in the preparation of monocyclic dilactams, the products are invariably contaminated with triethylamine hydrochloride, and the yields of the pure products are lower.

The same sequence of reactions are used to prepare macrobicyclic diamines with polyether links.[5,6] The reduction of macrobicyclic

$$N\underset{CH_2CH_2-O-CH_2CH_2-O-CH_2CH_2}{\overset{CH_2CH_2-O-CH_2CH_2-O-CH_2CH_2}{\diagdown}}N$$

diamides can be accomplished with diborane using the procedure described by Brown and Heim[7], but the present procedure is less involved and gives good yields.

1. Central Research Department, E. I. Du Pont de Nemours and Co., Wilmington, Delaware 19898.
2. A. Müller and L. Kindlman, *Chem. Ber.* **74B**, 416 (1941).
3. H. Stetter and J. Marx, *Justus Liebigs Ann. Chem.*, **607**, 59 (1957).
4. H. E. Simmons and C. H. Park, *J. Amer. Chem. Soc.*, **90**, 2428 (1968).
5. H. E. Simmons and C. H. Park, unpublished work.
6. B. Dietrich, J. M. Lehn, and J. P. Sauvage, *Tetrahedron Lett.*, 2885 (1969).
7. H. C. Brown and P. Heim, *J. Amer. Chem. Soc.*, **86**, 3566 (1964).

ENDOCYCLIC ENAMINE SYNTHESIS: N-METHYL-2-PHENYL-Δ²-TETRAHYDROPYRIDINE

$$C_6H_5COCH_3 + CH_3NH_2 \xrightarrow[(C_2H_5)_2O, C_6H_{14}, -30°]{TiCl_4} C_6H_5\overset{\overset{NCH_3}{\|}}{C}\!=\!CH_3$$

$$C_6H_5\overset{\overset{NCH_3}{\|}}{C}CH_3 + BrCH_2CH_2CH_2Cl \xrightarrow[\text{tetrahydrofuran}, -30°]{LiN(i\text{-}C_3H_7)_2} C_6H_5\overset{\overset{NCH_3}{\|}}{C}(CH_2)_3CH_2Cl$$

$$C_6H_5\overset{\overset{NCH_3}{\|}}{C}(CH_2)_3CH_2Cl \xrightarrow{\text{reflux}} \underset{\underset{CH_3}{|}}{\underset{N}{C_6H_5}}\!\!\!\diagup\!\!\!\bigcirc$$

Submitted by D. A. Evans[1] and L. A. Domeier
Checked by R. Decorzant and G. Büchi

1. Procedure[2]

A. *N-(α-Methylbenzylidene)-methylamine.* Into a dry, premarked, nitrogen-purged, 1-l. flask (Note 1) equipped with a mechanical stirrer, dry ice–acetone condenser with drying tube, and a 250-ml. pressure-equalizing addition funnel topped by a gas inlet connection was condensed approximately 70 ml. of methylamine which was passed through a potassium hydroxide trap. The flask was cooled in a methanol–ice bath, and a solution of 48 g. (0.4 mole) of acetophenone (Note 2) in 200 ml. of dry ether (Note 3) was added through the addition funnel. The addition funnel was rinsed with 25 ml. of dry ether, purged with nitrogen, and charged with 220 ml. of 1*M* titanium tetrachloride solution in hexane (Note 4). This solution was added to the cooled reaction flask over a 1.5-hour period (Note 5). After stirring an additional 30 minutes in the methanol–ice bath and 30 minutes at room temperature, the mixture was filtered through a Büchner funnel into a 1-l., round-bottomed flask (Note 6), and the solid material was rinsed with an additional 100 ml. of ether. The solvents were removed on a rotary evaporator, and the yellow residue transferred to a 100-ml. round-bottomed flask. Distillation through a short, vacuum-jacketed, Vigreux column yielded 37–47 g. (70–88%) of the colorless imine, b.p. 93–95° (11 mm.) (Notes 7 and 8).

B. *N-Methyl-2-phenyl-Δ²-tetrahydropyridine.* Into a 500-ml., nitrogen-purged flask equipped with serum cap, reflux condenser with nitrogen inlet connection, thermometer, and stirring bar, was introduced 100 ml. of dry tetrahydrofuran (Note 9) and 21.0 ml. (0.155 mole) of diisopropylamine (Note 10). The solution was cooled to $-30°$ with an acetone bath to which dry ice was added as needed, and 72 ml. (0.155 mole) of 2.14M n-butyllithium in hexane (Note 11) was added while keeping the temperature below 0°. After cooling the mixture to $-40°$, 20.6 g. (0.155 mole) of the imine (from part A) was added via syringe over a period of about 2 minutes, and the resulting yellow solution maintained at $-40°$ to $-30°$ for 15 minutes and then cooled to $-60°$. To the cold solution was added 16.5 ml. (0.16 mole) of 1-bromo-3-chloropropane (Note 12) in one portion via syringe while the temperature was maintained below $-40°$. The reaction mixture was maintained between $-60°$ and $-50°$ for 5 minutes, the bath was removed, and the mixture allowed to warm to room temperature. The reaction mixture was refluxed 3 hours to effect final ring closure (Note 13). After the addition of 150 ml. of aqueous 10% potassium carbonate to the cooled solution, the reaction mixture was stirred several minutes and then transferred to a nitrogen-purged separatory funnel. The reaction flask was rinsed with 100 ml. of 1:1 benzene–ether which was added to the separatory funnel, and the entire mixture was diluted with another 150 ml. of water. After shaking, the aqueous layer was removed, and the organic layer was washed with 100 ml. of brine, shaken with anhydrous granular sodium sulfate, and filtered into a 1-l., round-bottomed flask. The solvents were removed on a rotary evaporator, and the residue transferred to a 100-ml., round-bottomed flask. Short-path distillation under high vacuum yielded 18.8–21.7 g. (70–81%) of pale yellow enamine, b.p. 87–88° (4 mm.) (Notes 14 and 15).

2. Notes

1. All three necks of the flask should be vertical and not set at an angle. This is to prevent the accumulation of large amounts of the methylamine complex of titanium tetrachloride on the sides of the reaction flask.

2. The acetophenone was purchased from Matheson, Coleman and Bell and used without further purification.

3. Anhydrous ether available from Mallinckrodt Chemical Co. can be used without further drying.

4. The titanium tetrachloride (purified grade) was purchased from J. T. Baker. A $1M$ titanium tetrachloride solution was prepared by diluting 55 ml. of titanium tetrachloride (1.73 g./ml.) to a volume of 500 ml. with hexane which had been passed through 40–50 g. of basic alumina (Activity I).

5. The addition funnel should be thoroughly, but gently, flushed with nitrogen before being charged with the titanium tetrachloride solution, and a slight flow of nitrogen should be maintained throughout the addition. This is to prevent the diffusion of methylamine into the funnel where it will form a red insoluble titanium tetrachloride–amine complex.

6. If done quickly, the filtration need not be done under nitrogen with no effect on the yield in the case of this particular imine.

7. Proton magnetic resonance (carbon tetrachloride) δ: 7.7 (multiplet, $2H$, aryl CH); 7.2 (multiplet, $3H$, aryl CH); 3.2 (singlet, $3H$, vinylic CH_3); 2.1 (singlet, $3H$, N—CH_3). Infrared (carbon tetrachloride) cm^{-1}: 1645, 1450, 1370, 1290. Purity was confirmed by gas chromatography on a 4-ft., 10% Carbowax $20M$ column at 165°.

8. The imine should be stored under nitrogen and exposed to the air as little as possible during handling.

9. The tetrahydrofuran was freshly distilled from lithium aluminum hydride. See *Org. Syn.*, **46**, 105 (1966) for a note concerning the hazards involved in purifying tetrahydrofuran.

10. The diisopropylamine was purchased from Aldrich Chemical Co. and was distilled from calcium hydride prior to use.

11. The n-butyllithium in hexane was purchased from Ventron Corp.

12. The 1-bromo-3-chloropropane was purchased from Matheson, Coleman and Bell and was distilled from phosphorus pentoxide prior to use.

13. If the mixture is not refluxed, the intermediate imine may be isolated.

14. Proton magnetic resonance (carbon tetrachloride) δ: 7.3 (multiplet, $5H$, aryl CH); 4.8 (triplet, $J = 4$ Hz., $1H$, vinylic CH); 3.0 (multiplet, $2H$, N—CH_2—); 2.4 (singlet, $3H$, N—CH_3); 2.1–1.5 (multiplet, $4H$, aliphatic CH). Infrared (carbon tetrachloride) cm^{-1}: 1640, 1605, 1500, 1460, 1375, 1360, 1130, 1040. Purity was confirmed by

gas chromatography on a 4-ft., 10% Carbowax 20 M column at 165°.

15. The enamine should be refrigerated under nitrogen and used within a few days.

3. Discussion

N-Methyl-2-phenyl-Δ^2-tetrahydropyridine and similar compounds have previously been prepared by the hydrolysis and decarboxylation of α-benzoyl-N-methyl-2-piperidone[3] and by the addition of phenyl Grignard reagents to N-methyl-2-piperidone followed by dehydration.[4] Both of these methods require that a heterocyclic ring already be present in the system. In contrast, this procedure offers a new flexible route to the construction of five- or six-membered heterocyclic rings which may easily be incorporated into a larger polycyclic product. Several examples[5] of this process that can be cited to demonstrate this utility are

A wide variety of more complex endocyclic enamines are thus made available as synthetic intermediates.

1. Department of Chemistry, University of California, Los Angeles, California 90024.
2. This procedure is based on the work of Weingarten: H. Weingarten, J. P. Chupp, and W. A. White, *J. Org. Chem.*, **32**, 3246 (1967).
3. K. H. Büchel, H. J. Schulze-Steinem, and F. Korte, U.S. Patent 3,247,213 (1966); K. H. Büchel and F. Korte, *Chem. Ber.*, **95**, 2438 (1962).
4. R. Lakes and O. Grossmann, *Coll. Czech. Chem. Commun.*, **8**, 533 (1936).
5. D. A. Evans, *J. Amer. Chem. Soc.*, **92**, 7593 (1970).

TRI-*tert*-BUTYLCYCLOPROPENYL FLUOROBORATE

(Cyclopropenylium, tri-*tert*-butyl tetrafluoroborate)

$(CH_3)_3CCH_2MgCl + (CH_3)_3CCH_2COCl \rightarrow$
$\qquad\qquad (CH_3)_3CCH_2COCH_2C(CH_3)_3 + MgCl_2$

$(CH_3)_3CCH_2COCH_2C(CH_3)_3 + 2\ Br_2 \rightarrow$
$\qquad\qquad (CH_3)_3CCHBrCOCHBrC(CH_3)_3 + 2\ HBr$

$(CH_3)_3CCHBrCOCHBrC(CH_3)_3 + 2\ (CH_3)_3COK \rightarrow$

[cyclopropenone structure with $(CH_3)_3C$ substituents] $+ 2\ (CH_3)_3COH + 2\ KBr$

[cyclopropenone] $\xrightarrow{\text{1. }(CH_3)_3CLi;\ \text{2. }H_2O;\ \text{3. }HBF_4,\ Ac_2O}$ [tri-*tert*-butylcyclopropenyl cation] BF_4^-

Submitted by J. CIABATTONI,[1] E. C. NATHAN, A. E. FEIRING, and P. J. KOCIENSKI
Checked by L. M. LEICHTER and S. MASAMUNE

1. Procedure

A. *Dineopentyl Ketone.* A dry, 2-l., three-necked flask is fitted with a reflux condenser, a precision pressure-equalizing addition funnel, and a mechanical stirrer. A gas inlet tube at the top of the condenser is used to maintain a static nitrogen atmosphere in the reaction vessel throughout the reaction. The flask is charged with 900 ml. of anhydrous ether, 45 g. (1.85 g. atoms) of magnesium turnings, and 74.6 g. (0.70 mole) of 1-chloro-2,2-dimethylpropane (Note 1). The rapidly stirred mixture is heated to gentle reflux, and 156 g. (0.83 mole) of ethylene dibromide in 150 ml. of dry ether is added over a 12-hour period (Note 2). After addition is complete, the reaction mixture is refluxed for an additional 2 hours. The mixture is then cooled to 0–5° in an ice bath, and 71.7 g. (0.53 mole) of *tert*-butylacetyl chloride (Note 3) in 150 ml. of dry ether is added dropwise to the rapidly stirred Grignard reagent over a period of 1.5 hours (Note 4). After the addition is complete, the mixture is stirred at 0–5° for an additional 1.5 hours and then poured with stirring onto a mixture of 800 g. of cracked ice and 150 ml. of concentrated hydrochloric acid. The ether layer is separated and washed consecutively with 100 ml. each of water, aqueous 5% sodium carbonate solution, and finally aqueous saturated sodium chloride solution. After drying over anhydrous magnesium sulfate, filtration, and removal of solvent with a rotary evaporator, the yellow residual oil is distilled under reduced pressure to afford 81.2 g. (90%) of dineopentyl ketone as a colorless oil, b.p. 86–90° (22 mm.) (Notes 5 and 6).

B. *α,α'-Dibromodineopentyl Ketone. Caution! The dibromoketone, a highly volatile compound with lachrymatory properties, is a skin irritant which may induce allergic effects. Therefore, steps B and C should be performed in a well-ventilated hood. Rubber gloves should be worn.*

A 1-l., three-necked flask is fitted with a thermometer, an addition funnel, a magnetic stirring bar, and a gas exit tube, which is connected with Tygon tubing to a funnel inverted over a beaker of water for trapping hydrogen bromide. The flask is charged with 82 g. (0.48 mole) of dineopentyl ketone in 500 ml. of dichloromethane, and the solution is cooled to 0–5° in an ice bath. Over a period of 5 hours, 160 g. (1.0 mole) of bromine is added dropwise (Note 7), and after addition is complete the mixture is stirred an additional hour at 0–5°. The reaction mixture is then carefully transferred to a 1-l. separatory funnel, and the excess bromine is destroyed by extraction with 100 ml. of aqueous saturated sodium sulfite solution (Note 8). After washing with 100 ml. of aqueous 5% sodium bicarbonate solution and 100 ml. of aqueous saturated sodium chloride solution, the organic layer is dried over magnesium sulfate, filtered, and evaporated with a rotary evaporator. The yellow, crystalline residue is dissolved in 350 ml. of hot hexane and cooled in ice to give the dibromoketone as white needles which are collected by suction filtration, washed with 100 ml. of cold hexane (Note 9), and air dried in a well-ventilated hood to afford 83 g. of product. Concentration of the mother liquors provides two additional crops of crystals (40 g. and 12 g.) to afford a total yield of 135 g. (85%) of dibromoketone, m.p. 69–72° (Note 10).

C. *Di-tert-butylcyclopropenone. Caution! The same precautions described in Part B should be exercised in this step.*

To a dry 1-l., three-necked flask fitted with an efficient mechanical stirrer, a low temperature thermometer, and a solid addition assembly (Note 11) is added 97 g. (0.30 mole) of dibromoketone and 700 ml. of anhydrous tetrahydrofuran (Note 12). After the reaction vessel has been flushed with nitrogen, a static nitrogen atmosphere is maintained in the reaction vessel throughout the remainder of the reaction. The rapidly stirred solution is cooled to −70° in a dry ice–acetone bath, and then 80 g. (0.71 mole) of powdered potassium *tert*-butoxide (Note 13) is added over a 2-hour period. After addition is complete, the mixture is stirred an additional hour at −70°, and then 50 ml. of aqueous 10% hydrochloric acid solution is added dropwise. The cooling bath is then removed, and the mixture is allowed to warm to room temperature. The precipitated salts are filtered, washed with 100 ml. of tetrahydrofuran, and the filtrate and washing are combined. Most of the tetrahydrofuran is removed under reduced pressure, the nearly colorless residue is dissolved in 450 ml. of hexane (Note 14), and this solution is extracted twice with 100 ml. of water and once with 100 ml. of aqueous saturated sodium chloride solution. The organic layer is dried over magnesium sulfate, filtered, and concentrated with a rotary evaporator to leave a pale yellow oil which crystallizes upon brief standing. Sublimation of the crude product at 55° (1 mm.) provides 39–41 g. (79–83%) of the cyclopropenone, m.p. 61–63° (Notes 15 and 16).

D. *Tri-tert-butylcyclopropenyl Fluoroborate.* A dry, 500-ml., three-necked flask equipped with a magnetic stirring bar, a pressure-equalizing addition funnel, and a nitrogen inlet system is charged with 56 ml. of a 2.34M solution of commercial *tert*-butyllithium (0.13 mole) in pentane (Note 17) and cooled in an ice bath to 0°. A solution of 20.0 g. (0.12 mole) of di-*tert*-butylcyclopropenone in 200 ml. of pentane is added over 30 minutes, and then the ice bath is removed. The reaction mixture is stirred an additional 30 minutes to give a nearly colorless mixture which is poured with vigorous stirring into 150 ml. of water. The pentane layer is separated, washed with two 75-ml. portions of water, dried over magnesium sulfate, and filtered. Removal of solvent using a rotary evaporator leaves a pale yellow oil which is transferred to a 1-l. Erlenmeyer flask and diluted with 600 ml. of ether. After cooling the ether solution in an ice bath to 0°, a freshly prepared 10% solution of fluoroboric acid in acetic anhydride (Note 18) is added with rapid magnetic stirring. After the resulting suspension is stirred for 20 minutes the white precipitate is collected on a sintered glass funnel (medium porosity) under vacuum and washed thoroughly with 75-ml. portions of ether. The product is then dissolved in a minimal amount of boiling acetone (*ca.* 300 ml.) and cooled in a freezer (*ca.* −25°) to afford 15.3 g. of tri-*tert*-butylcyclopropenyl fluoroborate as white needles after filtration and washing with ether. Concentration of the mother liquors gives two further crops of 6.6 and 2.1 g. of pure product. The total yield is 24–28 g. (68–79%). When heated on a hot stage, the material darkens with decomposition above 300° (Note 19).

2. Notes

1. Reagent grade anhydrous ether is employed in all operations without prior purification. The magnesium turnings were available from the J. T. Baker Chemical Company, and the 1-chloro-2,2-dimethylpropane (b.p. 84–85°), obtained from Matheson, Coleman and Bell, should be distilled before use.

2. The ethylene dibromide (Eastman Organic Chemicals) is used without prior purification.

3. The *tert*-butylacetyl chloride (b.p. 126–129°) was purchased from Aldrich Chemical Company, Inc. and used without prior purification.

4. The inverse addition of the Grignard reagent to an ethereal solution of the *tert*-butylacetyl chloride does not improve the yield.

5. Fractional distillation is not necessary since the only volatile components are ether and dineopentyl ketone. A dark, high-boiling residue remains in the pot.

6. The proton magnetic resonance spectrum ($CDCl_3$ solution) of this product shows two singlets at δ 1.03 (18H) and 2.29 (4H). The infrared spectrum (CCl_4 solution) exhibits bands at 2950 (s), 2900 (s), 2865 (s), 1715 (s), 1480 (s), 1470 (s), 1370 (s), and 1355 cm^{-1} (s).

7. Irradiation with a sun lamp or addition of one drop of concentrated hydrochloric acid may be necessary to initiate the bromination.

8. *Caution! The extraction with sodium sulfite should be performed with caution since a considerable amount of heat is evolved.*

9. The dibromoketone is quite soluble in hexane; therefore, filtration should be conducted as rapidly as possible. The hexane should be precooled to at least 0°.

10. Very pure dibromoketone (m.p. 70–71°) may be obtained by sublimation at 25° (1 mm.). However, the product obtained by recrystallization is sufficiently pure for the next step. The proton magnetic resonance spectrum ($CDCl_3$ solution) shows singlets at δ 1.18 (18*H*) and 4.42 (2*H*). The infrared spectrum (CCl_4 solution) shows peaks at 2960 (s), 2935 (m), 2910 (m), 2870 (m), 1740 (s), 1720 (m), 1715 (m), 1485 (s), 1475 (s), 1400 (m), 1375 (s), and 1340 $cm.^{-1}$ (s).

11. A convenient apparatus for the addition of potassium *tert*-butoxide consists of a 250-ml. filter flask connected to the reaction vessel with Gooch tubing. The side arm of the filter flask serves as a nitrogen inlet.

12. The tetrahydrofuran should be freshly distilled from lithium aluminum hydride.

13. Alcohol-free potassium *tert*-butoxide, obtained from the MSA Research Corporation, Callery, Pennsylvania, should be weighed and transferred under anhydrous conditions.

14. Hexane is most effective in permitting extraction of residual tetrahydrofuran into water. Failure to remove the tetrahydrofuran can delay the crystallization of the cyclopropenone.

15. The checkers noted the presence of an oil which also distilled onto the cold-finger during the sublimation. The easiest way to remove this oil is to press the sublimed cyclopropenone between two sheets of filter paper. The melting point is recorded after having removed the oil in this manner. However, this oil poses no hindrance in the next step. Alternatively, the submitters report that pure cyclopropenone can be obtained more rapidly but in slightly reduced yield by recrystallization of the crude product from pentane at low temperature (*ca.* $-70°$).

16. The proton magnetic resonance spectrum ($CDCl_3$ solution) shows a singlet at δ 1.34. The infrared spectrum (CCl_4 solution) shows bands at 2980 (s), 2945 (m), 2920 (m), 2880 (m), 1875 (m), 1855 (s), 1820 (s), 1640 (s), 1485 (m), 1465 (m), and 1375 cm.$^{-1}$ (m). The ultraviolet spectrum exhibits a maximum at 260 mμ (log ϵ 1.66) in 95% ethanol and at 285 mμ (log ϵ 1.79) in cyclohexane. The mass spectrum shows peaks at *m/e* 166, 138, 123, 95, 81, and 67.

17. The *tert*-butyllithium was obtained from Alfa Inorganics, Inc.

18. *Caution! The reaction of fluoroboric acid with acetic anhydride is exothermic and should be conducted with caution.* Under an atmosphere of nitrogen, 102 g. (1.0 mole) of acetic anhydride (J. T. Baker Chemical Company) is cooled to $-40°$ in a dry ice–acetone bath. Then, with magnetic stirring, 20.4 g. (0.12 mole) of 50% fluoroboric acid (J. T. Baker Chemical Company) is added over 10 minutes. After carefully warming to $0°$, the freshly prepared solution is used immediately.

19. The proton magnetic resonance spectrum ($CDCl_3$ solution) shows a singlet at δ 1.58. The infrared spectrum (KBr) exhibits bands at 2980 (s), 1485 (s), 1465 (m), 1425 (m), 1370 (s), 1225 (s), 1197 (m), 1070 (s), 940 (m), 860 (m), and 525 cm.$^{-1}$ (m).

3. Discussion

The preparation of tri-*tert*-butylcyclopropenyl fluoroborate involves a modification of the method originally employed in the synthesis of triphenylcyclopropenyl perchlorate[2] from diphenylcyclopropenone.[2,3] The overall yields of tri-*tert*-butylcyclopropenyl fluoroborate and di-*tert*-butylcyclopropenone are 40% and 52%, respectively. Other substituted di-*tert*-butylcyclopropenyl cations have also been prepared by the above procedure.[4] It should be pointed out, however, that the synthesis of cyclopropenyl cations using the reaction of cyclopropenones with organometallic reagents is not general.[2,4] Possible competing side reactions include conjugate addition[5] and proton abstraction (if the cyclopropenone contains an acidic hydrogen).

1. Department of Chemistry, Brown University, Providence, Rhode Island 02912.
2. R. Breslow, T. Eicher, A. Krebs, R. A. Peterson, and J. Posner, *J. Amer. Chem. Soc.*, **87,** 1320 (1965).
3. R. Breslow and J. Posner, *Org. Syn.*, **47,** 62 (1967).
4. J. Ciabattoni and E. C. Nathan, III, *J. Amer. Chem. Soc.*, **91,** 4766 (1969).
5. J. Ciabattoni, P. J. Kocienski, and G. Melloni, *Tetrahedron Lett.*, 1883 (1969).

CUMULATIVE AUTHOR INDEX

This index comprises the names of contributors to Volume 50, 51, 52, 53, and 54. A number in boldface type denotes the volume, a number in ordinary type indicates the page of that volume

Ackerman, J.H. **53**, 52
Albertson, Noel F. **51**, 100
Andrews, S.D. **50**, 27
Arsenijevic, L. **53**, 5
Arsenijevic, V. **53**, 5
Auerbach, Robert A. **54**, 49
Aumiller, James C. **51**, 55, 73, 106

Backlawski, Leona M. **52**, 75
Bailey, Denis M. **51**, 100
Beck, A.K. **51**, 39, 76
Benkeser, Robert A. **50**, 88
Berlin, K.D. **53**, 98
Beverung W.N. **51**, 103
Bey, P. **53**, 63
Birkofer, L. **50**, 107
Bogdanowicz, Mitchell J. **54**, 27
Borch, Richard F. **52**, 124
Breuer, A. **54**, 11
Bridson, John N. **53**, 35
Brinkmeyer, R.S. **54**, 42
Brown, H.C. **52**, 59
Brun, P. **51**, 60

Cadogan, M. **51**, 109
Caglioti, L. **52**, 122
Calder, A. **52**, 77
Calzada, Jose G. **54**, 63
Carabateas, C.D. **53**, 52
Carlson, H.D. **50**, 81
Carlson, Robert M. **50**, 24
Carpino, Louis A. **50**, 31, 65
Castro, C.E. **52**, 62, 128
Chabala, John C. **54**, 60
Chan, Yihlin **53**, 48
Ciabattoni, J. **54**, 97
Cleland, George H. **51**, 1

Collington, E.W. **54**, 42
Collins, J.C. **52**, 5
Colon, Ismael **52**, 33
Connor, D.S. **52**, 16
Cookson, R.C. **51**, 121
Corey, E.J. **50**, 72
Crandall, J.K. **53**, 17
Crawley, L.C. **53**, 17
Creger, P.L. **50**, 58
Crowther, G.P. **51**, 96
Crumrine, David S. **54**, 49
Curphey, T.J. **51**, 142

Dawson, D.J. **54**, 71, 74, 77
Day, A.C. **50**, 3, 27
Dehesch, T. **54**, 84
DeMaster, Robert D. **52**, 135
Domeier, L.A. **54**, 93
Dorsey, James E. **50**, 15
Dowd, S.R. **54**, 46
Doyle, T.W. **50**, 13
DuBois, Richard H. **51**, 4
Dueber, Thomas E. **54**, 79
Duncan, J.H. **51**, 66

Eliel, E.L. **50**, 13, 38
Ellison, David L. **54**, 49
Epstein, William W. **53**, 48
Erickson, Bruce W. **54**, 19
Evans, D.A. **54**, 93

Feiring, A.E. **54**, 97
Feutrill, G.I. **53**, 90
Fishel, D.L. **50**, 102
Forrester, A.R. **52**, 77
Freeman, Fillmore **51**, 4

Gale, D.M. **50**, 6

CUMULATIVE AUTHOR INDEX

Gall, Martin 52, 39
Gassman, P.G. 53, 38
Gates, John W., Jr. 51, 82
Gault, R. 51, 103
Geluk, H.S. 53, 8
Gilbert, E.C. 50, 13
Greenlee, W.J. 53, 38
Gregson, Michael 54, 68
Grieco, Paul A. 54, 68
Griffin, Gary W. 52, 33
Gund, Tamara M. 53, 30
Gunsher, J. 51, 60
Gupte, S.S. 51, 121
Gurien, H. 51, 8

Hampton, K. Gerald 51, 128
Hanack, M. 54, 84
Harris, Thomas M. 51, 128; 53, 56
Hassner, A. 51, 53, 112
Hatch, Lewis F. 50, 43
Hauser, Charles R. 51, 90, 96, 128
Hawks, George H. 52, 36
Hay, J.V. 53, 56
Hayase, Yoshio 53, 44, 104
Heaney, Harry 54, 58
Heathcock, C.H. 51, 53, 112
Heck, R.F. 51, 17
Helquist, Paul M. 52, 115
Henery, J. 50, 21, 36
Hepburn, S.P. 52, 77
Hess, W.W. 52, 5
Hesse, A. 50, 66
Hetzel, Frederick W. 51, 139
Heyne, Hans-Ulrich 52, 11
Hill, D.T. 51, 60
Hill, Richard K. 50, 24
Hirata, Yoshimasa 53, 86
Ho, Beng T. 51, 136
Hooz, John 53, 35, 77; 54, 63
Horeau, A. 53, 5
House, Herbert O. 52, 39; 54, 49
Howes, P.D. 53, 1
Hummel, K. 53, 84
Hutchins, R.O. 50, 13, 38; 53, 21, 107

Ide, Junya 50, 62

Insalaco, Michael A. 50, 9
Ireland, Robert E. 53, 63, 116; 54, 71, 74, 77
Iwai, Issei 50, 62

Jacques, J. 53, 5, 111
Jefford, C.W. 51, 60
Jerkunica, J.M. 53, 94
Johnson, M. Ross 51, 11
Johnson, Robert E. 51, 100
Jones, E. 50, 75, 104

Kaiser, Carl 50, 94; 51, 48
Kaiser, Edwin M. 50, 88; 51, 96
Karlsson, Sune M. 53, 25
Keizer, V.G. 53, 8
Klein, G.W. 52, 16
Klug, W. 54, 11
Knoeber, M., Sr. 50, 38
Kocienski, P.J. 54, 97
Kofron, William G. 52, 75
Kono, Hiromichi 53, 77
Krimen, Lewis I. 50, 1

Lampman, Gary M. 51, 55, 73, 106
Larsen, A.A. 53, 52
Larsson, Lars-Eric 53, 25
Lee, Ving 50, 77
Le Gras, J. 51, 60
Ley, Steven V. 54, 58
Liedhegener, Annemarie 51, 86
Lohaus, G. 50, 18, 52

Mao, Chung-Ling 51, 90
Marolewski, T.A. 52, 132
Marquet, A. 53, 111
Martin, E.L. 51, 70
Maryanoff, Bruce E. 53, 21, 107
Mathai, I.M. 50, 97
McAdams, Louis V., III 50, 31, 65
McDonald, Richard N. 50, 15, 50
McKillop, Alexander 52, 36
McLaughlin, Thomas G. 51, 4
McMurry, John E. 53, 59, 70
Meinwald, J. 52, 53
Merényi, Ferenc 52, 1
Meyers, A.I. 51, 24, 103; 54, 42
Middleton, W.J. 50, 6, 81
Milewski, Cynthia A. 53, 107

Miller, Sidney I. 50, 97
Mirrington, R.N. 53, 90
Moodie, I.M. 50, 75, 104
Morris, Don L. 53, 98
Morrison, W.H., III 51, 31
Muchmore, D.C. 52, 109
Murakami, M. 52, 96
Muslinger, Walter J. 51, 82

Nagata, Wataru 52, 90, 96, 100; 53, 44, 104
Nathan, E.C. 54, 97
Newkome, G.R. 50, 102
Newman, Melvin S. 50, 77; 51, 139
Nierth, A. 54, 84
Nilsson, Martin 52, 1
Niznik, G.E. 51, 31
Noland, Wayland E. 52, 135

O'Connell, E.J., Jr. 52, 33
Odom, H.C. 51, 115
Ohme, Roland 52, 11
Olmstead, M.P. 50, 94
Owsley, D.C. 52, 128

Pachter, I.J. 54, 33, 37, 39
Park, Chung Ho 54, 88
Pearce, P.J. 52, 19
Pedersen, Charles J. 52, 66
Pettit, R. 50, 21, 36
Pigott, Foster 52, 83
Pinder, A.R. 51, 115
Politzer, Ieva R. 51, 24
Preuschhof, Helmut 52, 11

Rachlin, A.I. 51, 8
Ragnarsson, Ulf 53, 25
Rappe, C. 53, 123
Rathke, Michael W. 53, 66
Raymond, P. 50, 27
Regitz, Manfred 51, 86
Reineke, Charles E. 50, 50
Richards, D.H. 52, 19
Rickborn, Bruce 51, 11
Rifi, M.R. 52, 22
Rubottom, George M. 54, 60
Rüter, Jörn 51, 86
Rydon, H.N. 51, 44

Sample, Thomas E., Jr. 50, 43
Sandberg, Bengt E. 53, 25
Scheinbaum, Monte L. 54, 33, 37, 39
Schleyer, Paul v. R. 53, 30
Scilly, N.F. 52, 19
Seebach, D. 50, 72; 51, 39, 76
Seifert, Wolfgang K. 50, 84
Selman, L.H. 51, 109
Semmelhack, Martin F. 52, 115
Shapiro, Robert H. 51, 66
Sheng, M.N. 50, 56
Siggins, J.E. 53, 52
Simmons, Howard E. 54, 88
Sims, James J. 51, 109
Smith, Claibourne D. 51, 133
Snider, Theodore E. 53, 98
Sondheimer, F. 54, 1
Srivastava, K.C. 53, 98
Stang, Peter J. 54, 79
Staskun, B. 51, 20
Steppel, Richard N. 50, 15
Stevens, I.D.R. 51, 121
Stirling, C.J.M. 53, 1
Stöckel, K. 54, 1
Stork, G. 54, 46, 68
Suzuki, H. 51, 94

Taniguchi, H. 50, 97
Tarbell, D. Stanley 50, 9
Taylor, Edward C. 52, 36
Taylor, G.N. 52, 16
Teuber, H.-J. 52, 88
Thielecke, Wilfried 53, 30
Toda, Masaaki 53, 86
Traylor, T.G. 53, 94
Trost, Barry M. 54, 27
Trust, Ronald I. 53, 116

van Es., T. 51, 20
Vogel, E. 54, 11

Wade, Ruth S. 52, 62
Wadsworth, Donald H. 53, 13
Waegell, B. 51, 60
Wagner, D.P. 51, 8
Wakabayashi, Toshio 53, 44, 104
Walborsky, H.M. 51, 31
Walker, K.E. 51, 136
Watts, C.T. 51, 121

Wegner, P. 50, 107
Wehrli, Pius A. 52, 83
Weinstock, Joseph 50, 94; 51, 48
Wheeler, Thomas N. 52, 53
Whiting, M.C. 50, 3, 27
Wittig, G. 50, 66
Woodruff, R.A. 51, 96

Woodward, R.B. 54, 33, 37, 39

Yamamura, Shosuke 53, 86
Yang, N.C. 52, 132
Yoshioka, M. 52, 90, 96, 100

Zajacek, J.G. 50, 56
Zweifel, G. 52, 59

CUMULATIVE SUBJECT INDEX

This index comprises material from Volumes 50, 51, 52, 53, and 54 only; for subjects of previous volumes see Collective Volumes I through V.

Entries in CAPITAL LETTERS refer to the titles of individual preparations. Entries in ordinary type letters refer to principal products and major by-products, special reagents or intermediates (which may or may not be isolated), compounds mentioned in the text, Notes or Discussions as having been prepared by the method given, and apparatus described in detail or illustrated by a figure. Numbers in **boldface** type denote the volumes. Numbers in ordinary type indicate pages on which a compound or subject is mentioned in the indicated volume

Acetaldehyde, directed condensation with benzophenone, **50**, 67
 reaction with cyclohexylamine, **50**, 67
Acetaldehyde cyclohexylidene-, **53**, 104
Acetals, synthesis of, **53**, 135, 139
Acetamide, N,N-dimethyl-, dimethylacetal, **54**, 77
Acetates, by oxymercuration, **53**, 96
Acetic acid, 1-hydroxycyclohexyl-, ethyl ester, **53**, 67
Acetic acid, trifluoro-, sodium salt, **54**, 86
Acetic anhydride, with 2-heptanone to give 3-n-butyl-2,4-pentanedione, **51**, 90
ACETIC FORMIC ANHYDRIDE, **50**, 1
Acetone azine, **50**, 2
ACETONE HYDRAZONE, **50**, 2, 28
Acetophenone, **54**, 93
 as sensitizer for irradiation of bicyclo[2.2.1]hepta-2,5-diene to give quadricyclane, **51**, 133

Acetophenone, 2-diazo, **53**, 35
Acetophenone, N,N-dimethylhydrazone, **50**, 102
ACETOPHENONE HYDRAZONE, **50**, 102
3β-ACETOXY-5α-CYANOCHOLESTANE-7-ONE, **52**, 100
1-Acetoxy-2-methylcyclohexene, **52**, 40
Acetylation, selective, of 2-methoxynaphthalene, **53**, 7
p-ACETYL-α-BROMOHYDROCINNAMIC ACID, **51**, 1
Acetyl chloride, reaction with sodium formate, **50**, 1
 with propylene, aluminum chloride, and quinoline to give *trans*-3-pentene-2-one, **51**, 116
Acetyl chloride *tert*-butyl-, **54**, 97
2-Acetylcycloheptane-1,3-dione, **52**, 4
2-ACETYLCYCLOPENTANE-1,3-DIONE, **52**, 1
2-Acetylcyclopentanone, from cyclopentanone and acetic anhydride, **51**, 93
2-Acetyl-4,4-dimethylcyclopen-

tane-1,3-dione, **52**, 4
2-Acetyl-5,5-dimethylcyclopentane-1,3-dione, **52**, 4
3-ACETYL-2,4-DIMETHYLFURAN, **53**, 1
Acetylenedicarboxylic acid, dimethyl ester, **50**, 25, 36
Acetylenes, reaction with trimethylsilyl azide, **50**, 109
Acetylenic sulfonium salts, furans from, **53**, 3
2-Acetylindane-1,3-dione, **52**, 4
2-ACETYL-6-METHOXYNAPHTHALENE, **53**, 5
2-Acetyl-4-methylcyclopentane-1,3-dione, **52**, 4
Acid anhydride, mixed, with sodium azide to give phenylcyclo pentanecarboxylic acid azide, **51**, 48
Acid chlorides, reduction to aldehydes, **53**, 55
Acid chlorides, aromatic, diazoketones from, **53**, 37
Acrylic acid, with p-acetylbenzenediazonium bromide, **51**, 1
Acylation, of enol esters, **52**, 1
Adamantane, oxidation of, **53**, 8, 12
ADAMANTANONE, **53**, 8,
Alcohols, alkyl bromides and chlorides from, **54**, 66
by cis-hydration of olefins, **53**, 84
from olefins, **53**, 94
from organoboranes, **53**, 77
hindered, esterification of, **51**, 98
primary, from terminal olefins, **53**, 85
oxidation of, **52**, 5
Aldehydes, alkylation of, **54**, 46
by reduction of acid chlorides, **53**, 55
from diazoketones and organoboranes, **53**, 82
β-hydroxyesters from, **53**, 69

preparation from primary alcohols, **52**, 5
using acetic formic anhydride, **50**, 2
using 1,3-dithiane, **50**, 74
reaction with 1,3-bis(methylthio)allyllithium, **54**, 27
with trimethylsily azide, **50**, 109
ALDEHYDES BY OXIDATION OF TERMINAL OLEFINS WITH CHROMYL CHLORIDE: 2,4,4-TRIMETHYLPENTANAL, **51**, 4
ALDEHYDES FROM ACID CHLORIDES BY MODIFIED ROSENMUND REDUCTION: 3,4,5-TRIMETHOXYBENZALDEHYDE, **51**, 8
ALDEHYDES FROM ACID CHLORIDES BY REDUCTION OF ESTER MESYLATES WITH SODIUM BOROHYDRIDE: CYCLOBUTANECARBOXALDEHYDE, **51**, 11
ALDEHYDES FROM ALLYLIC ALCOHOLS AND PHENYLPALLADIUM ACETATE: 2-METHYL-3-PHENYLPROPIONALDEHYDE, **51**, 17
ALDEHYDES FROM AROMATIC NITRILES: p-FORMYLBENZENESULFONAMIDE, **51**, 20
ALDEHYDES FROM 2-BENZYL-4,4,6-TRIMETHYL-5,6-DIHYDRO-1,3-(4H)-OXAZINE: 1-PHENYLCYCLOPENTANECARBOXYALDEHYDE, **51**, 24
ALDEHYDES FROM 4,4-DIMETHYL-2-OXAZOLINE AND GRIGNARD REAGENTS: o-ANISALDEHYDE, **54**, 42
1-d-ALDEHYDES FROM ORGANOMETALLIC REAGENTS: 1-d-2-METHYLBUTANAL, **51**, 31
ALDEHYDES FROM PRIMARY ALCOHOLS BY OXIDATION WITH CHROMIUM TRIOXIDE: 1-HEPTANAL, **52**, 5
ALDEHYDES FROM sym-TRITHIANE: n-PENTADECANAL, **51**, 39
Aldehydes, acetylenic, **54**, 45
Aldehydes, aromatic, **54**, 45
Aldehydes, benzyl, **54**, 45
Aldehydes, olefinic, **54**, 45

CUMULATIVE SUBJECT INDEX

Aldehydes, α-phenyl-, from 2-benzyl-4,4,6-trimethyl-5,6-dihydro-1,3(4H)-oxazine, 51, 29
Aldehydes, α,β-unsaturated-, 53, 104
  δ-acetoxy-, 54, 26
  γ-hydroxy-, 54, 19
Aldehydes-1-d, 54, 45
Aldehydes-1-^{14}C, 54, 45
Aldimines, preparation of, 54, 46, 48
Aldol condensation, 54, 49
ALDOL CONDENSATIONS, DIRECTED, 50, 66
Alkenals, 54, 45
Alkylation, by oxonium salts, 51, 144
  intramolecular to form cyclopropanes, 52, 35
  of acids, 50, 61
  of lithium enolates, 52, 39
  of malonitrile, 53, 24
  with alkyl halide-silver salt complex, 54, 31
  with benzyl chloromethyl ether, 52, 17
N-Alkylation, in dimethyl sulfoxide, 54, 58
  in hexamethylphosphoramide, 54, 60
ALKYLATIONS OF ALDEHYDES *via* REACTION OF THE MAGNESIO-ENAMINE SALT OF AN ALDEHYDE: 2,2-DIMETHYL-3-PHENYLPROPIONALDEHYDE, 54, 46
Alkylboranes, oxidation, 52, 59
  synthesis, 52, 59
Alkyl bromides, from alcohols, 54, 66
  from alcohols, benzyl bromide, and triphenyl phosphite, 51, 47
Alkyl chlorides, from alcohols, 54, 63, 68
  from alcohols, benzyl chloride, and triphenyl phosphite, 51, 47
N-ALKYLINDOLES FROM THE ALKYLATION OF INDOLE SODIUM IN HEXAMETHYLPHOSPHORAMIDE: 1-BENZYLINDOLE, 54, 60
ALKYL IODIDES: NEOPENTYL IODIDE, IODOCYCLOHEXANE, 51, 44
Alkynals, 54, 45
Alkynols, cyclobutanones from, 54, 84
Allenylacetylenes, 50, 101
Allylic alcohols, allylic chlorides from, 54, 63, 68, 70
  by rearrangement of epoxides, 53, 20
ALLYLIC CHLORIDES FROM ALLYLIC ALCOHOLS: GERANYL CHLORIDE, 54, 68
π-Allylnickel bromide, 52, 199
Aluminum amalgam, 52, 78
Aluminum chloride, with ethylene and p-methoxyphenylacetyl chloride to give 6-methoxy-β-tetralone, 51, 109
  with propylene and acetyl chloride to give 4-chloropentan-2-one, 51, 116
  with succinic anhydride and isopropenyl acetate, 52, 1
Amides, by oxymercuraction, 53, 96
Amides, macrobicyclic, reduction of, 54, 89
Amine oxides, anhydrous, 50, 55, 58
Amines, by oxymercuration, 53, 96
  by reduction of nitriles, 53, 24
  protecting group for, 50, 12
AMINES FROM MIXED CARBOXYLIC-CARBONIC ANHYDRIDES: 1-PHENYLCYCLOPENTYLAMINE, 51, 48
p-Aminoacetophenone, diazotization of, 51, 1
Amino acids, N-*tert*-butyloxycarbonyl-, 53, 28
tert-Amyl iodide, from *tert*-amyl alcohol, methyl iodide, and triphenyl phosphite, 51, 47

ANDROSTAN-17β-OL, 52, 122
o-Anisaldehyde, preparation of, 54, 42
Annelation, isoxazole use in, 53, 75
[14]Annulene, 54, 9
[16]Annulene, 54, 9
[18]ANNULENE, 54, 1
[18]Annulene, tridehydro, 54, 3
[20]Annulene, 54, 9
[22]Annulene, 54, 9
[24]Annulene, 54, 9
[24]Annulene, tetradehydro, 54, 4
Anthracene, cyanation, 50, 55
Arndt-Eistert reaction, modified, 50, 77
γ-Aryl-β-diketones, general synthesis, 51, 131
Axial alcohols, preparative methods, 50, 15
AZETIDINE, 53, 13
Azetidine, 1-(2-carbethoxyethyl)-, 53, 13
AZIDOFORMIC ACID, tert-BUTYL ESTER, 50, 9
AZIRIDINES FROM β-IODOCARBAMATES: 1,2,3,4-TETRAHYDRO NAPHTHALENE(1,2)IMINE, 51, 53
Azoalkanes, synthesis, 52, 11
Azo-n-butane, 52, 15
Azocyclohexane, 52, 15
AZOETHANE, 52, 11
2-Azo-2-methylpropane, 52, 15
Azo-p-nitrobenzene, 52, 15
Azo-n-propane, 52, 15

BASE-INDUCED REARRANGEMENT OF EPOXIDES TO ALLYLIC ALCOHOLS: trans-Pinocarveol, 53, 17
Benzaldehyde, by condensation of phenyllithium with 1,1,3,3-tetramethylbutylisonitrile, 51, 38
 by reduction of benzonitrile with Raney nickel alloy, 51, 22
Benzaldehyde, 3,5-dinitro-, 54, 52

Benzaldehyde, 2-methoxy-, 54, 42
BENZALDEHYDE, 3,4,5-TRIMETHOXY-, 51, 8
BENZENESULFONAMIDE, p-FORMYL-, 51, 20
Benzhydrol, 52, 22
Benzoic acid-O-d, 53, 41
BENZONITRILE, 2,4-DIMETHOXY-, 50, 52
Benzophenone, directed reaction with acetaldehyde, 50, 68
1,4-Benzoquinone, 2,3-dichloro-5,6-dicyano-, (DDQ), aromatization with, 54, 14
Benzoylacetone, from acetophenone and acetic anhydride, 51, 93
BENZYL CHLOROMETHYL ETHER, 52, 16
1-BENZYLINDOLE, 54, 58, 60
2-BENZYL-2-METHYLCYCLOHEXANONE, 52, 39
2-BENZYL-6-METHYLCYCLOHEXANONE, 52, 39
3-Benzyloxy-4,5-dimethoxybenzaldehyde, by reduction of 3-benzyloxy-4,5-dimethoxybenzoyl chloride, 51, 10
2-Benzyl-4,4,6-trimethyl-5,6-dihydro-1,3(4H)-oxazine, from 2-methyl-2,4-pentanediol and phenylacetonitrile, 51, 27
Bicyclic diamines with bridgehead nitrogen, 54, 92
BICYCLO[1.1.0]BUTANE, 51, 55; 52, 32
Bicyclo[2.2.1]hepta-2,5-diene, irradiation sensitized by acetophenone to give quadricyclane, 51, 133
Bicyclo[2.2.0]hexa-2,5-diene, 50, 51
exo-Bicyclo[2.2.0]hexan-2-ol, 50, 51
Bicyclo[2.2.0]hex-2-ene, 50, 51
Bicyclo[2.2.0]hex-5-ene-2,3-dicarboxylic anhydride, 50, 51

BICYCLO[3.2.1]OCTAN-3-ONE, 51, 60
Binor-S, 53, 30
Binor-S, tetrahydro-, 53, 31
BIPHENYL, 51, 82
Biphenyls, unsymmetrical, 50, 27
2,2'-Bipyridyl, use as an indicator for organolithium reagents, 52, 39, 112
1,1-BIS(BROMOMETHYL)CYCLOPROPANE, 52, 22
Bis(CHLOROMETHYL) ETHER, hazard note, 51, 148
Bis(trifluoromethyl)carbene, 50, 8
BIS(TRIFLUOROMETHYL)DIAZOMETHANE, 50, 6
Borane, bis(3-methyl-2-butyl)-, 53, 79
Borane, trihexyl-, 53, 77
Boranes, oxidation with H_2O_2, 50, 90; 52, 59
2-BORNENE, 51, 66
Boron trifluoride, with dimethyl ether and epichlorohydrin to give trimethyloxonium tetrafluoroborate, 51, 142
Boron trifluoride-acetic acid, with acetic anhydride and 2-heptanone to give 3-n-butyl-2,4-pentanedione, 51, 90
Boron trifluoride etherate, co-catalyst, 53, 30, 32
α-Bromination, selective of aralkyl ketone, 53, 111
Bromine, with 3-chlorocyclobutanecarboxylic acid and mercuric oxide to give 1-bromo-3-chlorocyclobutane, 51, 106
Bromoacetaldehyde dimethyl acetal, 52, 139
1-BROMO-3-CHLOROCYCLOBUTANE, 51, 106
1-Bromo-3-chlorocyclobutane, with sodium to give bicyclo[1.1.0]-butane, 51, 55

3-Bromocyclobutanecarboxylic acid, 51, 75
Bromocyclopropane, from cyclopropanecarboxylic acid, 51, 108
(2-Bromoethyl)benzene, 50, 59
2-Bromo-6'-methoxy-2'-acetonaphthone, 53, 111
3-(Bromoethyl)cyclobutyl bromide, from 3-(bromomethyl)cyclobutane-carboxylic acid, 51, 108
α-Bromophenylacetic acid, 50, 31
2-Bromothiophene, 50, 75
1,3-BUTADIENE-1,4-DIOL trans, trans-DIACETATE, 50, 24
1,3-BUTADIENE, 2,3-DIPHENYL-, 50, 62
BUTANAL, 1-d-2-METHYL-, 51, 31
erythro-2,3-Butanediol monomesylate, by reaction of trans-2-butene oxide with methanesulfonic acid, 51, 11
3,5,1,7-[1,2,3,4]Butanetetraylnaphthalene, decahydro-, 53, 30
tert-Butanol, with p-toluoyl chloride and butyllithium to give tert-butyl p-toluate, 51, 96
2-Butanone, 1,3-dibromo-, 53, 123
2-Butanone, 3-methyl-, 54, 79
1-Butene, 1,3-bis(methylthio)-, 54, 24
1-Butene, 4-chloro-, 53, 119
trans-2-Butene oxide, from trans-2-butene and peracetic acid, 51, 13
3-Buten-1-ol, 53, 120
tert-Butylamine, N-(2-methylpropylidene)-, preparation of, 54, 46
tert-BUTYL AZIDOFORMATE, 50, 9
tert-BUTYLCARBONIC DIETHYLPHOSPHORIC ANHYDRIDE, 50, 9
cis-4-tert-BUTYLCYCLOHEXANOL, 50, 13
4-tert-Butylcyclohexanone, 50, 13

1-tert-Butylcyclohexene, by reduction of tert-butylbenzene, **50**, 92
2-tert-BUTYL-1,3-DIAMINOPROPANE, **53**, 21
5-tert-Butyl-2,3-dimethyliodobenzene, from iodine and 4-tert-butyl-1,3-dimethylbenzene, **51**, 95
tert-Butyl hydroperoxide, **50**, 56
N-tert-Butylhydroxylamine, **52**, 78
tert-Butyl hypochlorite, with 4-phenylurazole to give 4-phenyl-1,2,4-triazoline-3,5-dione, **51**, 123
n-Butyllithium, **50**, 104; **52**, 19
  reaction with 2-benzyl-4,4,6-trimethyl-5,6-dihydro-1,3-(4H)-oxazine, **51**, 25
  with sym-trithiane and 1-bromotetradecane, **51**, 40
  with 1,3-dithiane and 1-bromo-3-chloropropane to give 5,9-dithiaspiro[3.5]nonane, **51**, 76
  with p-toluoyl chloride and tert-butanol to give tert-butyl p-toluate, **51**, 96
sec-Butyllithium, with 1,1,3,3-tetramethylbutyl isonitrile and deuterium oxide to give N-(1-d-2-methylbutylidene)-1,1,3,3-tetramethylbutylamine, **51**, 33
tert-BUTYLMALONIC ACID, DIETHYL ESTER, **50**, 38
3-BUTYL-2-METHYLHEPT-1-EN-3-OL, **52**, 19
tert-Butyloxycarbonyl group, protection of amino acids by, **53**, 27
tert-BUTYLOXYCARBONYL-L-PROLINE, **53**, 25
3-n-BUTYL-2,4-PENTANEDIONE, **51**, 90
tert-Butyl phenylacetate, from phenylacetyl chloride, tert-butanol, and n-butyllithium, **51**, 98
tert-Butyl pivalate, from pivaloyl chloride and tert-butanol, **51**, 98
n-Butyl sulfide, with tetracyanoethylene oxide to give carbonyl cyanide, **51**, 70
tert-BUTYL p-TOLUATE, **51**, 96
3-Butyn-1-ol, preparation of, **54**, 86
Butyraldehyde, **54**, 51
γ-Butyrolactones, **54**, 32

Camphor tosylhydrazone, with methyllithium to give 2-bornene, **51**, 66
Carbinol, diphenyl, 2-dimethylamino-5-methylphenyl-, **53**, 56
β-CARBOLINE, 1,2,3,4-TETRAHYDRO-, **51**, 136
Carbon dioxide, anhydrous, **50**, 9
Carbon tetrachloride-organophosphine, chlorides from alcohols with, **54**, 63
CARBONYL CYANIDE, **51**, 70
Carbonyl cyanide, with alcohols, **51**, 72
  with amines, **51**, 72
  with olefins, **51**, 72
Carboxylic acids, α,β-unsaturated, cis-, **53**, 126
4-(p-Carboxyphenyl)-4H-1,4-thiazine 1,1-dioxide, **52**, 127
Cations, cyclopropenyl, **54**, 97
o-Chlorobenzaldehyde, by reduction of o-chlorobenzonitrile with Raney nickel alloy in formic acid, **51**, 23
p-Chlorobenzaldehyde, by reduction of p-chlorobenzonitrile with Raney nickel alloy, **51**, 22
m-Chlorobenzoyl chloride, **50**, 16
3-Chlorobicyclo[3.2.1]oct-2-ene, from exo-3,4-dichloribicyclo[3.2.1]-oct-2-ene and lithium

aluminum hydride, 51, 61
with sulfuric acid to give bicyclo[3.2.1]octan-3-one, 51, 62
3-CHLOROCYCLOBUTANECARBOXYLIC ACID, 51, 73
3-Chlorocyclobutanecarboxylic acid, with mercuric oxide and bromine to give 1-bromo-3-chlorocyclobutane, 51, 106
3-Chloro-1,1-cyclobutanedicarboxylic acid, from sulfuryl chloride and 1,1-cyclobutanedicarboxylic acid, 51, 73
4-Chloropentan-2-one, with quinoline to give trans-3-penten-2-one, 51, 116
m-CHLOROPERBENZOIC ACID, 50, 15, 34
3-Chloropropionitrile, 50, 20
Chlorosulfonyl isocyanate, in nitrile synthesis, 50, 52
 precautions, 50, 18
2-Chloro-5-thiophenethiol, 50, 106
Cholestane, 53, 86
3α-Cholestanol, 50, 15
5β-CHOLEST-3-ENE-5-ACETALDEHYDE, 54, 71
5β-Cholest-3-ene-5-acetamide, N,N-dimethyl, 54, 77
5β-Cholest-3-ene-5-acetic acid, ethyl ester, 54, 74
4-Cholestene, 3-ethenyloxy-, 54, 71
4-Cholesten-3β-ol, 54, 71, 75, 77
Chromium(II)-en perchlorate, 52, 62
Chromium(II) salts, standardization procedure for solutions, 52, 64
Chromium trioxide-pyridine complex, 52, 5
Chromyl chloride, oxidation of terminal olefins, 51, 6
Cinnamaldehyde, by reduction of cinnamonitrile with Raney nickel alloy in formic acid, 51, 23
 from the ester-mesylate, 51, 16
Cinnamic acid, 50, 18
CINNAMONITRILE, 50, 18
Claisen rearrangement, 53, 121; 54, 73
Claisen-amide rearrangement, 54, 78
Claisen-ester rearrangement, 54, 74
Clemmensen reduction, modified, 53, 86
Cobalt(II) bromide-triphenylphosphine, catalyst, 53, 30, 32
Condensation of p-acetylbenzenediazonium bromide with acrylic acid, 51, 1
Conduritol-D, 50, 27
Conjugate addition of Grignard reagents, 50, 41
CONTROLLED POTENTIAL ELECTROLYTIC REDUCTION: 1,1-BIS(BROMOMETHYL)CYCLOPROPANE, 52, 22
Copper(I) chloride, reaction with an organomagnesium compound, 50, 98
 use in Grignard reactions, 50, 39
Copper(I) phenylacetylide, 52, 128
Coupling of acetylenes and halides, copper-promoted, 50, 100
m-Cresol, 5-methoxy-, 53, 90
(Z)-Crotonic acid, 53, 123
Crotonic acid, β-pyrrolidino-, ethyl ester, 53, 60
Crown polyethers, complexes, 52, 71, 73
 synthesis, 52, 66
Cupric acetate-pyridine, oxidative coupling with, 54, 2, 9
Cuprous oxide, in thiol arylation, 50, 75
Curtius reaction, modification using mixed carboxylic-

carbonic anhydrides, 51, 51
Cyanation of aromatic compounds, 50, 53
9-Cyanoanthracene, 50, 55
p-Cyanobenzenesulfonamide, reduction with Raney nickel alloy to p-formylbenzenesulfonamide, 51, 20
p-Cyano-N,N-diethylaniline, 50, 54
Cyanohydrins, formation by use of alkylaluminum cyanides, 52, 96
1-Cyano-1-hydroxy-6-methoxytetralin, 52, 96
Cyanomesitylene, 50, 54
1-CYANO-6-METHOXY-3,4-DIHYDRONAPHTHALENE, 52, 96
1-Cyano-2-methoxynaphthalene, 50, 55
4-Cyano-1-methoxynaphthalene, 50, 55
2-Cyanothiophene, 50, 54
N-Cyanovinylpyrrolidone, 50, 54
CYCLIC KETONES FROM 1,3-DITHIANE: CYCLOBUTANONE, 51, 76
Cyclization, solvolytic, 54, 84
Cycloalkene oxides, 1-methyl, conversion to exocyclic methylene alcohols, 53, 20
Cyclobutadiene, generation in situ, 50, 23
CYCLOBUTADIENEIRON TRICARBONYL, 50, 21
Cyclobutane, 52, 32
CYCLOBUTANE, 1-BROMO-3-CHLORO-, 51, 106
Cyclobutanecarbonyl chloride, reaction with erythro-2,3-butanediol monomesylate, 51, 12
CYCLOBUTANECARBOXALDEHYDE, 51, 11
CYCLOBUTANECARBOXYLIC ACID, 3-CHLORO-, 51, 73
1,1-Cyclobutanedicarboxylic acid, with sulfuryl chloride to give 3-chloro-1,1-cyclobutanedicarboxylic acid, 51, 73
CYCLOBUTANONE, 51, 76
CYCLOBUTANONE VIA SOLVOLYTIC CYCLIZATION, 54, 84
Cyclobutanones, 54, 32
CYCLOBUTENE, cis-3,4-DICHLORO-, 50, 36
2-Cyclobutyl-cis-4-trans-5-dimethyl-1,3-dioxolane, by reaction of erythro-3-methanesulfonyloxy-2-butyl cyclobutanecarboxylate with sodium borohydride, 51, 12
hydrolysis to cyclobutanecarboxaldehyde, 51, 13
3,5-CYCLOHEXADIENE-1,2-DICARBOXYLIC ACID, 50, 50
$\Delta^{1,\alpha}$-Cyclohexaneacetaldehyde, 53, 104
Cyclohexaneacetic acid, 1-hydroxy-, ethyl ester, 53, 66
Cyclohexane carbonitrile, 50, 20
1,4-Cyclohexanediol, from hydroquinone, 51, 105
Cyclohexanol, with triphenyl phosphite and methyl iodide, 51, 45
CYCLOHEXANOL, 4-tert-BUTYL, cis-, 50, 13
Cyclohexanol, 1-methyl-, 53, 94
Cyclohexanone, 54, 40
annelation to, 53, 75
CYCLOHEXANONE, 2-DIAZO-, 51, 86
Cyclohexanone, 2-(3-ethyl-5-methyl-4-isoxazolylmethyl)-, 53, 72
Cyclohexanone, 2-hydroxymethylene-, preparation of, 54, 38
reaction with alkylenedithiotosylates, 54, 37
Cyclohexanone, 2,2-trimethylenedithio-, 54, 39
4-CYCLOHEXENE-1,2-DICARBOXYLIC ACID, DIETHYL ESTER, trans-, 50, 43
2-Cyclohexen-1-one, 4,4-dimethyl-, 53, 48
2-Cyclohexenone, 5-isopropyl-2-methyl-, 53, 63

1-Cyclohexenylpyrrolidine, 54, 39
Cyclohexylamine, 52, 127
 reaction with acetaldehyde, 50, 67
Cyclohexylideneacetaldehyde, 53, 104
1,3,5,7,9,11,13,15,17-Cyclo-öctadecanonaene, 54, 1
Cyclooctatetraene, chlorination, 50, 36
 reaction with mercuric acetate, 50, 24
CYCLOOCTENE, 1-NITRO-, 50, 84
CYCLOPENTANECARBOXALDEHYDE, 1-PHENYL-, 51, 24
Cyclopentane-1,3-dione, 52, 4
Cyclopentanones, 54, 32
CYCLOPENTYLAMINE, 1-PHENYL-, 51, 48
Cyclopropane, 52, 32
Cyclopropanecarboxaldehyde, by reduction of ester-mesylate, 51, 16
CYCLOPROPANECARBOXYLIC ACID, cis-2-PHENYL-, 50, 94
Cyclopropane derivatives, synthesis, 52, 22, 33, 132
Cyclopropenes, 50, 30
Cyclopropenone, di-tert-butyl-, 54, 98
Cyclopropenylium, tri-tert-butyl, tetrafluoroborate, 54, 97
CYCLOPROPYLDIPHENYLSULFONIUM FLUOROBORATE, 54, 27

1-DECALOL, 51, 103
2-Decalol, dehydration, 50, 92
n-Decane, 53, 107
Decarboxylation, of 3-chloro-1,1-cyclobutanedicarboxylic acid to 3-chlorocyclobutanecarboxylic acid, 51, 74
Dehalogenation with sodium-liquid ammonia, 54, 13
Dehydrohalogenation, with quinoline, 51, 116
 with triethylamine, 52, 36
DEHYDROXYLATION OF PHENOLS:
HYDROGENOLYSIS OF PHENOLIC ETHERS: BIPHENYL, 51, 82
O-Demethylation, selective, of dimethyl ethers, 53, 93
Deoxygenation, selective, of ketones, 53, 89
Deuterium content determination in deuterodiazomethane, 53, 41
"Dewar benzene," 50, 51
trans-7,8-Diacetoxybicyclo[4.2.0]octa-2,4-diene, 50, 24
trans, trans-1,4-DIACETOXY-1,3-BUTADIENE, 50, 24
N,N'-Dialkylsulfamides, synthesis, 52, 11
DIAMANTANE: PENTACYCLO[7.3.1.1^{4,12}.0^{2,7}.0^{6,11}]TETRADECANE, 53, 30
1,2-DIAROYLCYCLOPROPANES: trans trans-1,2-DIBENZOYLCYCLOPROPANE, 52, 33
1,10-Diazacyclooctadecane, 54, 88
1,10-Diazacyclooctadecane-2,9-dione, 54, 88
DIAZOACETOPHENONE, 53, 35
1-(Diazoacetyl)naphthalene, 50, 77
Diazoalkanes, hydrogen-deuterium exchange in, 53, 43
16-Diazoandrost-5-en-3β-ol-17-one, 52, 54
2-DIAZOCYCLOALKANONES: 2-DIAZOCYCLOHEXANONE, 51, 86
2-Diazocycloalkanones, from α-(hydroxymethylene)ketones with p-toluenesulfonyl azide, 51, 88
2-DIAZOCYCLOHEXANONE, 51, 86
α-Diazoketones, reaction with trialkylborane, 53, 82
 rearrangement, 52, 53
 synthesis, 52, 53
α-Diazoketones, aromatic, from acid chlorides, 53, 37
α-Diazoketones, cyclic, alkylation of, 53, 82, 83
Diazomethane, in modified Arndt-

Eistert reaction, 50, 77
preparation and hazard note, 53, 35, 36, 38, 39
spectrophotometric determination, 53, 40
DIAZOMETHANE, BIS(TRIFLUOROMETHYL)-, 50, 6
Diazomethane, dideuterio-, 53, 38
2-DIAZOPROPANE, 50, 5, 27
DIBENZO-18-CROWN-6 POLYETHER, 52, 66
trans-1,2-DIBENZOYLCYCLOPROPANE, 52, 33
Dibenzyl sulfide, 50, 33
Diborane, 50, 90; 52, 59
 reduction of dinitrile with, 53, 24
α,α'-DIBROMODIBENZYL SULFONE, 50, 31, 65
2,2-Dibromo-6'-methoxy-2'-acetonaphthone, 53, 111
1,3-Dicarbonyl compounds, furans from, 53, 3
exo-3,4-Dichlorobicyclo[3.2.1]oct-2-ene from norbornene and ethyl trichloroacetate, 51, 60
1,4-Dichlorobutadiene, 50, 37
Dichlorocarbene-isotetraline addition products, 54, 13
cis-3,4-DICHLOROCYCLOBUTENE, 50, 21, 36
DICYCLOHEXYL-18-CROWN-6-POLYETHER, 52, 66
DIDEUTERIODIAZOMETHANE, 53, 38
Diels-Alder adduct, pyrolysis, 50, 37
Diels-Alder reaction, 50, 37
 of 1,4-diacetoxy-1,3-butadiene, 50, 27
 using 3-sulfone, 50, 47
2,6-Diethoxy-1,4-oxathiane, 52, 135
DIETHYLALUMINUM CYANIDE, 52, 90
N,N-Diethylaniline, cyanation, 50, 54
DIETHYL tert-BUTYLMALONATE, 50, 38

Diethyl carbonate, with hydrazine hydrate to give ethyl hydrazinecarboxylate, 51, 121
DIETHYL 2-(CYCLOHEXYLAMINO)VINYLPHOSPHONATE, 53, 44
Diethyl 2,2-diethoxyethylphosphonate, 53, 44
Diethyl formylmethylphosphonate, 53, 45
Diethyl fumarate, as a dineophile, 50, 43
Diethyl isopropylidenemalonate, 50, 38
Diethyl malonate, condensation with acetone, 50, 39
Diethyl 5-methylcoprost-3-enyl phosphate, 52, 109
Diethyl phosphorochloridate, 50, 10
 reaction with metal enolates, 52, 109
N,N'-Diethylsulfamide, 52, 11
DIETHYL trans-Δ^4-TETRAHYDROPHTHALATE, 50, 43
Dihydrocarvone, 53, 63
1,2-Dihydronaphthalene, with iodine isocyanate and methanol to give methyl (trans-2-iodo-1-tetralin)carbamate, 51, 112
trans-1,2-DIHYDROPHTHALIC ACID, 50, 50
cis-1,2-Dihydrophthalic anhydride, 50, 51
Diimines, macrocyclic, 54, 88
Diiododurene, from durene and iodine, 51, 95
Diiron enneacarbonyl, 50, 21
Diketones, from diazoketones and organoboranes, 53, 82
β-DIKETONES FROM METHYL ALKYL KETONES: 3-n-BUTYL-2,4-PENTANEDIONE, 51, 90
2,6-Dimethoxybenzaldehyde, by reduction of 2,6-dimethoxybenzonitrile with Raney nickel alloy in formic acid, 51, 23
2,4-DIMETHOXYBENZONITRILE, 50, 52

Dimethylamine, (1,1-dimethoxyethyl)-, 54, 77
3,4-Dimethylbenzaldehyde, by reduction of 3,4-dimethylbenzoyl chloride, 51, 10
4,5-DIMETHYL-1,2-BENZOQUINONE, 52, 88
1,3-Dimethylbicyclo[1.1.0]butane, 52, 32
N,N-DIMETHYL-5-β-CHOLEST-3-ENE-5-ACETAMIDE, 54, 77
4,4-DIMETHYL-2-CYCLOHEXEN-1-ONE, 53, 48
N,N-DIMETHYLCYCLOHEXYLAMINE, 52, 124
1,2-Dimethylcyclopropane, 52, 32
N,N-Dimethyldodecylamine, 50, 56
N,N-DIMETHYLDODECYLAMINE OXIDE, 50, 56
Dimethyl ether, with boron trifluoride diethyl etherate and epichlorohydrin to give trimethyloxonium tetrafluoroborate, 51, 142
2,4-Dimethyl-3-furyl methyl ketone, 53, 1
N,N-Dimethylhydrazine, 50, 102
4,4-Dimethyl-2-neopentylpentanal, by oxidation of 4,4-dimethyl-2-neopentyl-1-pentene with chromyl chloride, 51, 6
2,2-Dimethyl-4-phenylbutyric acid, 50, 58
Dimethylprop-2-ynylsulfonium bromide, preparation, intermediate in furan synthesis, 53, 1
2,4-Dimethyl-3-sulfolene, in Diels-Alder reaction, 50, 48
3,4-Dimethyl-3-sulfolene, in Diels-Alder reaction, 50, 48
Dimethyl sulfoxide, N-alkylation in, 54, 58
Dimethyl sulfoxide, sodium salt, 50, 62

Dimethylthiocarbamyl chloride, synthesis of, 51, 140
with 2-naphthol to give O-2-naphthyl dimethylthiocarbamate, 51, 139
3,5-DINITROBENZALDEHYDE, 53, 52
Dinitrogen tetroxide, 50, 84
Diphenylacetyl chloride, 52, 36
Diphenylacetylene, conversion to diphenylbutadiene, 50, 63
2,3-DIPHENYL-1,3-BUTADIENE, 50, 62
2,2-Diphenylethanal, by oxidation of 1,1-diphenylethylene with chromyl chloride, 51, 6
2,2-Diphenylethyl benzoate, from 2,2-diphenylethanol, benzoyl chloride, and n-butyllithium, 51, 98
Diphenyliodonium chloride, with 2,4-pentanedione and sodium amide to give 1-phenyl-2,4-pentanedione, 51, 128
DIPHENYLKETENE, 52, 36
α,α'-Diphenylthiodiglycolic acid, 50, 31
2,3-DIPHENYLVINYLENE SULFONE, 50, 32, 34, 65
Dipotassium nitrosodisulfonate, 52, 86, 88
Dipyridine chromium(VI) oxide, 52, 5
DIRECTED ALDOL CONDENSATIONS: threo-4-HYDROXY-3-PHENYL-2-HEPTANONE, 54, 49
DIRECTED LITHIATION OF AROMATIC COMPOUNDS: (2-DIMETHYLAMINO-5-METHYLPHENYL)DIPHENYLCARBINOL, 53, 56
DIRECT IODINATION OF POLYALKYLBENZENES: IODODURENE, 51, 94
Disiamylborane, 53, 79
Disodium hydroxylaminedisulfonate, 52, 83
DISODIUM NITROSODISULFONATE, 52, 83
1,3-DITHIANE, 50, 72

1,3-Dithiane, with 1-bromo-3-chloropropane and n-butyllithium to give 5,9-dithiaspiro[3.5]nonane, **51**, 76
1,3-Dithianes, preparation of, **54**, 39
1,4-Dithiaspiro[4.5]decan-6-one, **54**, 37
5,9-Dithiaspiro[3.5]nonane, from 1,3-dithiane, 1-bromo-3-chloropropane, and n-butyllithium, **51**, 76
1,5-Dithiaspiro[5.5]undecan-7-one, **54**, 39
2,2'-DITHIENYL SULFIDE, **50**, 75
1,3-Dithiolanes, **54**, 37
Diynes, preparation, **50**, 101
n-Dodecane, **53**, 108
Double bond, exocyclic, selective hydrogenation, **53**, 65
DURENE, IODO-, **51**, 94

Electrolytic reduction, apparatus, **52**, 23
Enamines, reaction with alkylenebisthiotosylates, **54**, 39
Enamines, N-bromomagnesium-, alkylation of, **54**, 46
Enamines endocyclic, synthesis of, **54**, 93, 96
ENDOCYCLIC ENAMINE SYNTHESIS: N-METHYL-2-PHENYL-Δ^2-TETRAHYDROPYRIDINE, **54**, 93
Enol acetates, acylation of, **52**, 1
Enolates, lithium salts, aldol condensation with, **54**, 49
Enol esters, preparation, **52**, 39
Epichlorohydrin, **54**, 20
  with boron trifluoride diethyl etherate and dimethyl ether to give trimethyloxonium tetrafluoroborate, **51**, 142
Epoxides, reaction with 1,3-bis(methylthio)allyllithium, **54**, 26, 27
  rearrangement to allylic alcohols, **53**, 20

ESTERIFICATION OF HINDERED ALCOHOLS: tert-BUTYL p-TOLUATE, **51**, 96
Esters, from diazoketones and organoboranes, **53**, 82
Esters, α-deuterio-, **53**, 82
Esters, γ,δ-unsaturated by Claisen rearrangement, **53**, 122
1,2-Ethanedithiol, di-p-toluenesulfonate, **54**, 33
Ethers, by oxymercuration, **53**, 96
Ethers, aryl methyl-, selective monodemethylation of, **53**, 93
Ethoxyethene, **54**, 71
ETHYL 5β-CHOLEST-3-ENE-5-ACETATE, **54**, 74
Ethyl diazoacetate, as source of carbethoxycarbene, **50**, 94
Ethylene, with p-methoxyphenylacetyl chloride and aluminum chloride to give 6-methoxy-β-tetralone, **51**, 109
Ethylenediamine, complexes with chromium(II) salts, **52**, 62
2,2-(ETHYLENEDITHIO)CYCLOHEXANONE, **54**, 37
Ethylene dithiotosylate, **54**, 33, 37
Ethyl hydrazinecarboxylate, from hydrazine hydrate and diethyl carbonate, **51**, 121
Ethyl 1-hydroxycyclohexylacetate, **53**, 67
Ethylidenecyclohexylamine, **50**, 66
Ethyl 1-iodopropionate, from ethyl 1-hydroxypropionate, methyl iodide, and triphenyl phosphite, **51**, 47
Ethyl 4-methyl-E-4,8-nonadienoate, preparation and apparatus for, **53**, 118
ETHYL 6-METHYLPYRIDINE-2-ACETATE, **52**, 75
Ethyl 1-naphthylacetate, **50**, 77

ETHYL PYRROLE-2-CARBOXYLATE, 51, 100
Ethyl trichloroacetate, with norbornene to give exo-3,4-dichlorobicyclo[3.2.1]oct-2-ene, 51, 60
Ethyl vinyl ether, 54, 71
Exocyclic methylene alcohols, from 1-methylcycloalkene oxides, 53, 20
Extractor, 54, 90

Favorskii rearrangement, 53, 127
Fluoroboric acid, sodium salt, preparation of, 54, 30
Fluoroboric acid-acetic anhydride, 54, 99
FORMATION AND ALKYLATION OF SPECIFIC ENOLATE ANIONS FROM AN UNSYMMETRICAL KETONE: 2-BENZYL-2-METHYLCYCLOHEXANONE AND 2-BENZYL-6-METHYLCYCLOHEXANONE, 52, 39
FORMATION AND PHOTOCHEMICAL WOLFF REARRANGEMENT OF CYCLIC α-DIAZO KETONES: D-NORANDROST-5-EN-3β-OL-16-CARBOXYLIC ACIDS, 52, 53
FORMIC ACID, AZIDO-, tert-BUTYL ESTER, 50, 9
Formylation, with acetic formic anhydride, 50, 2
p-FORMYLBENZENESULFONAMIDE, 51, 20
Formyl fluoride, 50, 2
Fremy's salt, 52, 86, 88
Furan, 3-acetyl-2,4-dimethyl-, 53, 1
Furans, from acetylenic sulfonium salts and 1,3-dicarbonyl compounds, 53, 3

GENERAL SYNTHESIS OF 4-ISOXAZOLECARBOXYLIC ESTERS: ETHYL 3-ETHYL-5-METHYL-4-ISOXAZOLECARBOXYLATE, 53, 59
Geraniol, 54, 63
GERANYL CHLORIDE, 54, 63, 68

Glyoxylic acid, with tryptamine to give 1,2,3,4-tetrahydro-β-carboline, 51, 136
Grignard reagent, aldehydes from, 54, 42
reaction with aldimines, ketimines, 54, 48
Grignard reagent, 2,2-dimethylpropylmagnesium chloride, 54, 97

Halides, reduction of, 53, 109
1,6-Heptadien-3-ol, 2-methyl- 53, 116
1-HEPTANAL, 52, 5
2-Heptanone, with acetic anhydride, boron trifluoride-acetic acid, and p-toluenesulfonic acid to give 3-n-butyl-2,4-pentanedione, 51, 90
2-Heptanone, threo-4-hydroxy-3-phenyl-, 54, 51
N-Heterocyclics, 5- and 6-membered, 54, 96
2,4-Hexadienenitrile, 50, 20
1,5-Hexadiyne, oxidative coupling of, 54, 1
Hexafluoroacetone hydrazone, 50, 6
HEXAFLUOROACETONE IMINE, 50, 6, 81
Hexafluorothioacetone, 50, 83
Hexamethylbicyclo[1.1.0]butane, from 1,3-dibromohexamethylcyclo butane and sodium-potassium alloy, 51, 58
Hexamethylphosphoramide, 50, 61
N-alkylation in, 54, 60
n-Hexanal, from 2-lithio-1,3,5-trithiane and 1-bromopentane, 51, 43
Hexanal, 2,2-dimethyl-5-oxo-, 53, 50
1-Hexanol, 53, 79
2-Hexenal, trans-4-hydroxy-, 54, 21
2,4-dinitrophenylhydrazone, 54, 105
4-Hexenal, 6-chloro-4-methyl-,

dimethyl acetal, **54,** 70
4-Hexenal, 6-hydroxy-4-methyl-, dimethyl acetal, **54,** 70
1-Hexen-4-ol, 1,3-bis(methylthio)-, **54,** 21
HOMOGENEOUS CATALYTIC HYDROGENATION: DIHYDROCARVONE, **53,** 63
Hunsdiecker reaction, modified; for preparation of 1-bromo-3-chlorocyclobutane, **51,** 106
Hydrazine, anhydrous, **50,** 3, 4, 6
  reaction with hydrazones, **50,** 102
Hydrazine hydrate, **50,** 3
Hydrazoic acid, safe substitute for, **50,** 107
HYDRAZONES, PREPARATION, **50,** 102
Hydroboration, of 2-methyl-2-butene, **50,** 90
HYDROBORATION OF OLEFINS: (+)-ISOPINOCAMPHEOL, **52,** 59
Hydroboration-oxidation of olefins, **53,** 83
HYDROCINNAMIC ACID, p-ACETYL-α-BROMO-, **51,** 1
Hydrocyanation, with alkylaluminum cyanides, **52,** 100
HYDROGENATION OF AROMATIC NUCLEI: 1-DECALOL, **51,** 103
Hydrogenation, catalytic homogeneous, **53,** 64
Hydrogen cyanide, reaction with triethylaluminum, **52,** 90, 100
Hydrogen-deuterium exchange in diazoalkanes, **53,** 43
HYDROGENOLYSIS OF CARBON-HALOGEN BONDS WITH CHROMIUM(II)-EN PERCHLORATE: NAPHTHALENE FROM 1-BROMONAPHTHALENE, **52,** 62
Hydrogenolysis, of phenolic ethers to aromatics, **51,** 85
  of p-(1-phenyl-5-tetrazolyloxy)biphenyl with palladium-on-charcoal catalyst to biphenyl, **51,** 83

Hydrolysis, of 5,9-dithiaspiro[3.5]nonane to cyclobutanone, **51,** 77
  of substituted sym-trithianes to aldehydes, **51,** 42
Hydroquinone, 2,3-dichloro-5,6-dicyano-, **54,** 14
3-Hydroxycyclohexanecarboxylic acid, from 3-hydroxybenzoic acid, **51,** 105
β-HYDROXY ESTERS FROM ETHYL ACETATE AND ALDEHYDES OR KETONES: ETHYL 1-HYDROXYCYCLOHEXYLACETATE, **53,** 66
2-(Hydroxymethylene)cyclohexanone with p-toluenesulfonyl azide to give 2-diazocyclohexanone, **51,** 86
γ-HYDROXY-α,β-UNSATURATED ALDEHYDES VIA 1,3-BIS(METHYLTHIO)ALLYLLITHIUM: $trans$-4-HYDROXY-2-HEXENAL, **54,** 19

Imines of haloketones, **50,** 83
Iminodipropionate, N-(3-chloropropyl), diethyl ester, **53,** 14
Iminodipropionate, N-(3-hydroxypropyl)-, diethyl ester, **53,** 13
Indole, sodium salt, preparation of, **54,** 60
Indole, 1-benzyl, **54,** 50, 58
Indoles, N-alkyl-, **54,** 58, 60
Iodides, from alcohols, methyl iodide, and triphenyl phosphite, **51,** 47
Iodine isocyanate, from silver isocyanate and iodine, **51,** 112
IODOCYCLOHEXANE, **51,** 45
IODODURENE, **51,** 94
$trans$-β-Iodoisocyanates, general synthesis from olefins with iodine isocyanate, **51,** 114
Iodometric titration, **50,** 17
Ion-exchange resin, Amberlite IR-120, cyclization with, **53,** 50

Iridium tetrachloride, in modified Meerwein-Ponndorf reduction, 50, 13
Iron enneacarbonyl, see Diiron enneacarbonyl
Irradiation apparatus, 51, 133
Irradiation, of bicyclo[2.2.1]hepta-2,5-diene to give quadricyclane, 51, 133
N-Isobutylaniline, 53, 127
Isobutyric acid, alkylation, 50, 59
Isocrotonic acid, 53, 123
(+)-ISOPINOCAMPHEOL, 52, 59
Isopropenyl acetate, acylation of, 52, 1
Isotetralin, 54, 11
ISOXAZOLE ANNELATION REACTION: 1-METHYL-4,4a,5,6,7,8-HEXAHYDRONAPHTHALEN-2(3H)-ONE, 53, 70
4-Isoxazolecarboxylic acid ester, 3,5-disubstituted, 53, 61
Isoxazole, 4-chloromethyl-3-ethyl-5-methyl-, 53, 71
Isoxazole, 3-ethyl-4-hydroxymethyl-5-methyl-, 53, 70

Ketimines, enamines from, in ketones alkylation, 54, 48
Ketone, aralkyl, selective α-bromination of, 53, 111
Ketone, α,α'-dibromodineopentyl-, preparation of, 54, 98
Ketone dineopentyl-, preparation of, 54, 97
Ketone, heptyl phenyl, 53, 78
Ketones, alkylation of, 54, 48
  Clemmensen reduction of, 53, 89
  conversion to α,β-unsaturated aldehydes, 53, 106
  homologous, from α-diazoketone and organoboranes, 53, 82
  β-hydroxyesters from, 53, 69
  α-deuterio-, 53, 82
  preparation using 1,3-dithiane, 50, 74; 51, 80
  reaction with 1,3-bis(methylthio)allyllithium, 54, 27
  trifluoromethanesulfonates from, 54, 83
KETONES AND ALCOHOLS FROM ORGANOBORANES: 1. PHENYL HEPTYL KETONE: 2. 1-HEXANOL; 3. 1-OCTANOL, 53, 77

Lead tetraacetate, oxidation of a hydrazone to a diazo compound, 50, 7
Lithio ethyl acetate, 53, 67
Lithium, reductions in amine solvents, 50, 89
Lithium aluminum hydride, reduction of exo-3,4-dichlorobicyclo-[3.2.1]oct-2-ene to 3-chlorobicyclo[3.2.1]oct-2-ene, 51, 61
Lithium aluminum tri-tert-butoxyhydride, 53, 53
Lithium amide, bis(trimethylsilyl), 53, 66
Lithium amide, diisopropyl-, 50, 67; 52, 43; 54, 21, 49, 94
Lithium, 1,3-bis(methylthio)allyl-, preparation of, 54, 21
Lithium, butyl-, 54, 21
Lithium, dimethylcuprate, 50, 41; 52, 109
Lithium, enolates, formation of, 52, 109
  preparaion and alkylation, 52, 39
  reaction with diethyl phosphorochloridate, 52, 109

MACROCYCLIC DIIMINES: 1,10-DIAZACYCLOÖCTADECANE, 54, 88
MACROCYCLIC POLYETHERS: DIBENZO-18-CROWN-6-POLYETHER AND DICYCLOHEXYL-18-CROWN-6-POLYETHER, 52, 66
Malonaldehyde bis(diethyl acetal), 52, 139
Malonaldehydic acid diethyl

acetal, 52, 139
Malononitrile, alkylation of, 53, 24
Malononitrile, *tert*-butyl-, 53, 21
Meerwein reaction, preparation of *p*-acetyl-α-bromohydrocinnamic acid, 51, 1
Mercuric acetate, 54, 71
  reaction with cyclooctatetraene, 50, 24
Mercuric chloride, hydrolysis with, 54, 21
Mercuric oxide, use in oxidation of hydrazones, 50, 28
  with 3-chlorocyclobutanecarboxylic acid and bromine to give 1-bromo-3-chlorocyclobutane, 51, 106
MERCURIC OXIDE-MODIFIED HUNSDIECKER REACTION: 1-BROMO-3-CHLOROCYCLOBUTANE, 51, 106
Mesitylene, cyanation, 50, 54
METHALATION OF 2-METHYLPYRIDINE DERIVATIVES: ETHYL 6-METHYLPYRIDINE-2-ACETATE, 52, 75
Metalation, directed, 53, 59
Methallyl alcohol, with phenylmercuric acetate to yield 2-methyl-3-phenylpropionaldehyde, 51, 17
METHALLYLBENZENE, 52, 115
π-Methallylnickel bromide, 52, 115
Methane-d_2, diazo-, 53, 38
Methanesulfonic acid, esters, 54, 82
Methanesulfonic acid, trifluoro-, alkenyl esters, 54, 82
  anhydride, 54, 79
  3-butyn-1-yl ester, 54, 84
  3-methyl-1-buten-2-yl ester, 54, 80
  3-methyl-2-buten-2-yl ester, 54, 80
  1,2-dimethylpropenyl ester, 54, 79
  vinyl esters, 54, 82
erythro-3-Methanesulfonyloxy-2-butyl cyclobutanecarboxylate, by reaction of *erythro*-2,3-butanediol monomesylate with cyclobutanecarbonyl chloride, 51, 12
Methanethiol, 54, 19
1,6-METHANO[10]ANNULENE, 54, 11
Methanol, (2-dimethylamino-5-methylphenyl)diphenyl-, 53, 56
6'-Methoxy-2'-acetonaphthone, 53, 5
o-Methoxybenzaldehyde, 54, 42
p-Methoxybenzaldehyde, by reduction of *p*-methoxybenzonitrile with Raney nickel alloy, 51, 22
1-Methoxynaphthalene, cyanation, 50, 55
2-Methoxynaphthalene, cyanation, 50, 55
3-Methoxy-4-nitrobenzaldehyde, by reduction of 3-methoxy-4-nitro-benzoyl chloride, 51, 10
p-Methoxyphenylacetyl chloride, with ethylene and aluminum chloride to give 6-methoxy-β-tetralone, 51, 109
6-METHOXY-β-TETRALONE, 51, 109
Methylal, 50, 72
Methylamine, *N*-(α-methylbenzylidene)-, preparation of, 54, 93
1-*d*-2-METHYLBUTANAL, 51, 31
bis-(3-Methyl-2-butyl)borane, 50, 90
N-(1-*d*-2-Methylbutylidene)-1,1,3,3-tetramethylbutylamine, from *sec*-butyllithium, 1,1,3,3-tetramethylbutyl isonitrile, and deuterium oxide, 51, 33
  from *sec*-butylmagnesium bromide with 1,1,3,3-tetramethyl butyl isonitrile and deuterium oxide, 51, 35
5-METHYLCOPROST-3-ENE, 52, 109
3-Methylcyclohexene, from 2-methylcyclohexanone tosyl-

hydrazone and methyllithium, 51, 69
N-Methylcyclohexylamine, 52, 127
3-Methylcyclopentane-1,3-dione, 52, 4
Methylenecyclopropanes, 50, 30
3-Methylheptan-4-ol, 52, 22
Methyl iodide, with triphenyl phosphite and cyclohexanol, 51, 45
  with triphenyl phosphite and neopentyl alcohol, 51, 44
METHYL (trans-2-IODO-1-TETRALIN)CARBAMATE, 51, 112
Methyl (trans-2-iodo-1-tetralin)carbamate, with potassium hydroxide to give 1,2,3,4-tetrahydronaphthalene (1,2)imine, 51, 53
Methyllithium, with camphor tosylhydrazone to give 2-borene, 51, 66
  ether solution, 50, 69
  standardization procedure, 50, 69; 52, 46
Methylmagnesium iodide, 1,4-addition in the presence of Cu(I), 50, 39
2-Methyl-2-nitropropane, 52, 78
2-METHYL-2-NITROSOPROPANE AND ITS DIMER, 52, 77
Methyl D-Norandrost-5-en-3β-ol-16β-carboxylate, 52, 56
(S)-(-)-3-Methylpentanal, from 2-lithio-1,3,5-trithiane and (S)-(+)-1-iodo-2-methylbutane, 51, 43
3-Methyl-2,4-pentanedione, from butanone and acetic anhydride, 51, 93
N-Methyl-α-phenylethylamine, 52, 127
2-METHYL-3-PHENYLPROPIONALDEHYDE, 51, 17
3-Methyl-3-phenylpropionaldehyde, from crotyl alcohol and phenyl palladium acetate, 51, 19

N-Methylpiperidine, 52, 127
3-Methyl-3-sulfolene, in Diels-Alder reaction, 50, 48
MODIFIED CLEMMENSEN REDUCTION: CHOLESTANE, 53, 86

Naphthalene, sodium-liquid ammonia reduction of, 54, 11
1-NAPHTHALENEACETIC ACID, ETHYL ESTER, 50, 77
Naphthalene, 2-acetyl-6-methoxy-, 53, 5
Naphthalene, 2-bromoacetyl-6-methoxy-, 53, 112
1-NAPHTHALENECARBAMIC ACID, 1,2,3,4-TETRAHYDRO-2-IODO-, METHYL ESTER, 51, 112
Naphthalene-1-carbonitrile, 50, 20
2-Naphthalenecarboxyaldehyde, by reduction of 2-naphthalene-carbonitrile, 51, 22
Naphthalene, 2,2-dibromoacetyl-6-methoxy-, 53, 112
NAPHTHALENE(1,2)IMINE, 1,2,3,4-TETRAHYDRO-, 51, 53
NAPHTHALENE, OCTAHYDRO-, 50, 88
Naphthalene, 1,4,5,8-tetrahydro-, 54, 12
2-NAPHTHALENETHIOL, 51, 139
3H-Naphthalen-2-one, 4,4a,5,6,7,8-hexahydro, 1-methyl, 53, 70
1-Naphthol, hydrogenation to 1-decalol, 51, 103, 104
2-Naphthol, with dimethylthiocarbamyl chloride to give O-2-naphthyldimethylthiocarbamate, 51, 139
1-Naphthoyl chloride, 50, 79
1-Naphthylacetic acid, propyl ester, 50, 80
O-2-Naphthyl dimethylthiocarbamate, from 2-naphthol and dimethylthiocarbamyl chloride, 51, 139
  thermolysis to S-2-naphthyl dimethylthiocarbamate, 51, 140
S-2-Naphthyl dimethylthiocar-

bamate, hydrolysis with potassium hydroxide to 2-naphthalenethiol, 51, 140
Needle valve adapter, 54, 90
Neopentyl alcohol, with triphenyl phosphite and methyl iodide, 51, 44
NEOPENTYL IODIDE, 51, 44
Nerol, 54, 70
Neryl chloride, 54, 70
Nickel carbonyl, precautions for handling, 52, 117
 reaction with allyl halides, 52, 115
Nitriles, from carboxylic acids, 50, 20
 from diazoketones and organoboranes, 53, 82
Nitroacetaldehyde diethyl acetal, 52, 139
Nitro compounds, preparation, 50, 88
1-NITROCYCLOOCTENE, 50, 84
Nitrogen atmosphere, apparatus for maintaining, 52, 46
Nitrogen, purification, 50, 69
1-Nitro-1-octadecene, 50, 86
o-Nitrophenylacetaldehyde dimethyl acetal, 52, 139
4-p-Nitrophenyl-1,2,4-triazoline-3,5-dione, synthesis of, 51, 125
Nitrosation, of ketones, 52, 53
Nitroso compounds, synthesis, 52, 77
4,8-Nonadienoic acid, 4-methyl-, trans, ethyl ester, 53, 116
Nonan-5-ol, 52, 22
D-NORANDROST-5-EN-3β-OL-16-CARBOXYLIC ACIDS, 52, 53
Norbornene, with ethyl trichloroacetate to give exo-3,4-dichloro-bicyclo[3.2.1]oct-2-ene, 51, 60

2,6-Octadiene, E-1-chloro-3,7-dimethyl-, 54, 63, 68
$\Delta^{1,9}$-Octalin, 50, 89
$\Delta^{1,10}$-OCTALIN, 50, 88

Octamethylenediamine, 54, 88
n-Octanal, from 2-lithio-1,3,5-trithiane and 1-bromoheptane, 51, 43
1-Octanol, 53, 79
trans-2-Octnal, 54, 27
1-Octene, conversion to 1-octanol, 53, 79
Olefins, conversion to alcohols, 53, 94
 from tosylhydrazones and methyllithium, 51, 69
 hydroboration-oxidation of, 53, 83
 hydrogenation of, 53, 65
 synthesis, 52, 109, 115
 terminal, with chromyl chloride, 51, 6
Olefins trisubstituted, stereoselective synthesis of, 53, 121
Orcinol, dimethyl ether, 53, 90
ORCINOL MONOMETHYL ETHER, 53, 90
Organoboranes, ketones and alcohols from, 53, 77
Organolithium compounds, 53, 58
Organolithium reagents, preparation, 52, 21
 standardization procedure, 52, 46
Organophosphine-carbon tetrachloride, chlorides from alcohols with, 54, 63
Orthoacetic acid, ethyl ester, 54, 75
2-Oxazoline, 4,4-dimethyl-, preparation of and aldehydes from, 54, 42, 43
 preparation of methiodide salt, 54, 44
Oxidation, of primary alcohols to aldehydes, 52, 5
 of terminal olefins with chromyl chloride, 51, 6
 of 2,4,4-trimethyl-1-pentene with chromyl chloride, 51, 4
 with chromium trioxide-pyridine complex, 52, 5

with hydrogen peroxide, 52, 59
with the nitrosodisulfonate radical, 52, 83, 88
with ozone, 52, 135
with potassium permanganate, 52, 77
with sodium hypobromite, 52, 77
with sodium hypochlorite, 52, 11
OXIDATION WITH THE NITROSODISULFONATE RADICAL. 1. PREPARATION AND USE OF DISODIUM NITROSODISULFONATE: TRIMETHYL-p-BENZOQUINONE, 52, 83
OXIDATION WITH THE NITROSODISULFONATE RADICAL. II. USE OF DIPOTASSIUM NITROSODISULFONATE (FREMY'S SALT): 4,5-DIMETHYL-1,2-BENZOQUINONE, 52, 88
Oxidative coupling, 54, 1
Oximes, preparation, 50, 88
16-Oximinoandrost-5-en-3β-ol-17-one, 52, 53
Oximino ketones, synthesis, 52, 53
Oxygen, analysis for active, 50, 16
OXYMERCURATION-REDUCTION: ALCOHOLS FROM OLEFINS 1-METHYLCYCLOHEXANOL, 53, 94
Ozonides, reduction with sulfur dioxide, 52, 135

Palladium 10% - calcium carbonate catalyst, preparation of, 54, 8
Palladium-on-charcoal catalyst, biphenyl from p-(1-phenyl-5-tetrazolyloxy)biphenyl and hydrogen, 51, 83
Pentacyclo[7.3.1.1^{4,12}.0^{2,7}.0^{6,11}]tetradecane, 53, 30
n-PENTADECANAL, 51, 39
n-Pentadecanal dimethyl acetal, by methanolysis of 2-tetradecyl-sym-trithiane, 51, 40
1,3-PENTADIYNE, 1-PHENYL-, 50, 97
1,4-PENTADIYNE, 1-PHENYL-, 50, 97
n-Pentanal, by condensation of butyllithium with 1,1,3,3-tetramethylbutyl isonitrile, 51, 38
2,4-Pentanedione, with sodium amide and diphenyliodonium chloride to give 1-phenyl-2,4-pentanedione, 51, 128
2,4-Pentanedione, 3-alkyl-, 51, 93
2,4-PENTANEDIONE, 3-n-BUTYL-, 51, 90
2,4-PENTANEDIONE, 1-PHENYL-, 51, 128
3-Pentanol, 52, 22
$trans$-2-Pentenal, 4-hydroxy-4-methyl-, 54, 27
Pent-1-en-3-ol, 52, 22
$trans$-3-PENTEN-2-ONE, 51, 115
PERBENZOIC ACID, m-CHLORO-, 50, 15
Periodic acid dihydrate, with iodine and durene to give iododurene, 51, 94
Phenols, from aryl methyl ethers, 53, 93
Phenylacetaldehyde, from 2-lithio-1,3,5-trithiane and benzyl bromide, 51, 43
Phenylacetic acid, bromination, 50, 31
Phenylacetone, 54, 50
Phenylacetonitrile, 50, 20
Phenylacetylene, reaction with ethyl magnesium bromide, 50, 98
Phenylation, of β-diketones with diphenyliodonium chloride, 51, 131
of ketoesters with diphenyliodonium chloride, 51, 131
of nitroalkanes with diphenyliodonium chloride, 51, 132
PHENYLATION WITH DIPHENYL-

IODONIUM CHLORIDE: 1-PHENYL-2,4-PENTANEDIONE, 51, 128
4-Phenyl-1-carbethoxysemicarbazide, from ethyl hydrazinecarboxylate and phenyl isocyanate, 51, 122
  with potassium hydroxide to give 4-phenylurazole, 51, 122
1-Phenyl-5-chlorotetrazole, with p-phenylphenol to give p-(1-phenyl-5-tetrazolyloxy)biphenyl, 51, 82
β-PHENYLCINNAMALDEHYDE, 50, 65
1-PHENYLCYCLOPENTANECARBOXYALDEHYDE, 51, 24
1-Phenylcyclopentanecarboxylic acid, with ethyl chlorocarbonate to give mixed carboxylic-carbonic anhydride, 51, 48
1-PHENYLCYCLOPENTYLAMINE, 51, 48
Phenylcyclopentylamine, by hydrolysis of phenylcyclopentyl isocyanate, 51, 49
Phenylcyclopentyl isocyanate, by thermolysis of phenylcyclopentane-carboxylic acid azide, 51, 49
2-(1-Phenylcyclopentyl)-4,4,6-trimethyl-5,6-dihydro-1,3(4H)-oxazine, from 2-benzyl-4,4,6-trimethyl-5,6-dihydro-1,3(4H)-oxazine, 1,4-dibromobutane, and n-butyllithium, 51, 24
2-(1-Phenylcyclopentyl)-4,4,6-trimethyltetrahydro-1,3-oxazine, by reduction of 2-(1-phenylcyclopentyl)-4,4,6-trimethyl-5,6-dihydro-1,3(4H)-oxazine with sodium borohydride, 51, 25
α-Phenylcyclopropane, 52, 32
1-Phenylcyclopropanecarboxaldehyde, from 2-benzyl-4,4,6-trimethyl-5,6-dihydro-1,3(4H)-oxazine, 51, 29
cis-2-PHENYLCYCLOPROPANECARBOXYLIC ACID, 50, 94
trans-2-Phenylcyclopropanecarboxylic acid, 50, 96
α-Phenylethylamine, 52, 127
2-Phenylethyl iodide, from 2-phenylethanol, methyl iodide, and triphenyl phosphite, 51, 47
Phenylethynylmagnesium bromide, 50, 97
Phenyl isocyanate, with ethyl hydrazinecarboxylate to give 4-phenyl-1-carbethoxysemicarbazide, 51, 122
Phenylmagnesium bromide, o-methoxy, preparation of, 54, 44
Phenylmercuric acetate, with methallyl alcohol to yield 2-methyl-3-phenylpropionaldehyde, 51, 17
1-PHENYL-1,3-PENTADIYNE, 50, 97
1-PHENYL-1,4-PENTADIYNE, 50, 97
α-Phenylpentanal, from 2-benzyl-4,4,6-trimethyl-5,6-dihydro-1,3(4H)-oxazine, 51, 29
1-PHENYL-2,4-PENTANEDIONE, 51, 128
3-Phenyl-2,4-pentanedione, from phenylacetone and acetic anhydride, 51, 93
p-Phenylphenol, with 1-phenyl-5-chlorotetrazole to give phenolic ether, 51, 82
1-PHENYL-4-PHOSPHORINANONE, 53, 98
2-Phenylpropanal, by oxidation of 2-phenylpropene with chromyl chloride, 51, 6
3-Phenylpropanal, from allyl alcohol and phenylpalladium acetate, 51, 19
2-PHENYL[3,2-b]PYRIDINE, 52, 128
p-(1-Phenyl-5-tetrazolyloxy)biphenyl, from p-phenylphenol and 1-phenyl-5-chlorotetrazole, 51, 82

hydrogenation to biphenyl, 51, 83
4-PHENYL-1,2,4-TRIAZOLINE-3, 5-DIONE, 51, 121
4-Phenyl-1,2,4-triazoline-3,5-dione, reactions of, 51, 126
Phenyltrimethylammonium sulfomethylate, 53, 111
Phenyltrimethylammonium tribromide, selective bromination with, 53, 112, 114
4-Phenylurazole, from 4-phenyl-1-carbethoxysemicarbazide and potassium hydroxide, 51, 122
  with tert-butyl hypochlorite to give 4-phenyl-1,2,4-triazoline-3,5-dione, 51, 123
Phosphine, bis(2-cyanoethyl)phenyl-, 53, 98
Phosphine, tributyl, 54, 65
Phosphine, trioctyl, 54, 65
Phosphine, triphenyl, 54, 65
Phosphine, trisdimethylamino, 54, 65
Phosphinimines, 50, 109
1H-Phosphinolin-4-one, 2,3-dihydro, 1-phenyl, 53, 102
Phosphonic acid, 2-(cyclohexylamino)vinyl-, diethyl ester, 53, 44
Phosphonic acid, 2,2-diethoxyethyl-, diethyl ester, 53, 44
Phosphonic acid, formylmethyl-, diethyl ester, 53, 45
4-Phosphorinanone, 1-phenyl-, 53, 98
Phosphorin-3-carbonitrile, 4-amino-1-phenyl-1,2,5,6-tetrahydro-, 53, 99
Phosphorus heterocycles, 53, 102
Photochemical reactions, dissociation of trihalomethanes, 52, 132
  rearrangement, 52, 53
Phthalic acid, reduction, 50, 50

Phthalic acid, 3-nitro-, hazard note, 53, 129
α-Pinene oxide, preparation of, 53, 18
2(10)-Pinen-3α-ol, 53, 17
trans-Pinocarveol, 53, 17
Pivalaldehyde, by condensation of tert-butylmagnesium bromide with 1,1,3,3-tetramethylbutyl isonitrile, 51, 38
  by reduction of ester-mesylate, 51, 16
Pivalnotrile, 50, 20
Polyalkylbenzenes, with iodine to give iodo derivatives, 51, 95
Polycyclics by annelation, 53, 76
Potassium acetate complex with dicyclohexyl-18-crown-6-polyether, 52, 71
Potassium amide, 52, 75
Potassium azide, 50, 10
Potassium tert-butoxide, 52, 53
Potassium hydroxide complex with dicyclohexyl-18-crown-6-polyether, 52, 71
PREPARATION AND REDUCTIVE CLEAVAGE OF ENOL PHOSPHATES: 5-METHYLCOPROST-3-ENE, 52, 109
PREPARATION OF CYANO COMPOUNDS USING ALKYLALUMINUM INTERMEDIATES. III. 3β-ACETOXY-5α-CYANOCHOLESTAN-7-ONE, 52, 100
PREPARATION OF CYANO COMPOUNDS USING ALKYLALUMINUM INTERMEDIATES. II. 1-CYANO-6-METHOXY-3,4-DIHYDRONAPHTHALENE, 52, 96
PREPARATION OF CYANO COMPOUNDS USING ALKYLALUMINUM INTERMEDIATES. I. DIETHYLALUMINUM CYANIDE, 52, 90
PREPARATION OF α,β-UNSATURATED ALDEHYDES via THE WITTIG REACTION: CYCLOHEXYLIDENE-

ACETALDEHYDE, 53, 104
PREPARATION OF VINYL TRIFLUORO-
  METHANESULFONATES: 3-METHYL-
  2-BUTEN-2-YL TRIFLATE, 54,
  79
L-Proline, tert-butyloxycar-
  bonyl-, 53, 25
Propanal, 2,2-dimethyl-3-
  phenyl-, 54, 46
  2,4-dinitrophenylhydrazone,
  54, 48
Propane, 1,3-bis(methylthio)-2-
  methoxy-, 54, 20
Propane, 1-bromo-3-chloro-, 54,
  94
Propane, 1-chloro-2,2-dimethyl-,
  54, 97
Propane, 1-chloro-3-iodo-,
  preparation of, 54, 29
Propane, 1-(p-toluenesulfonyl)-
  3-(p-toluenethiosulfonyl)-,
  54, 34, 36
1,3-Propanediamine, 2-tert-
  butyl-, 53, 21
1,3-Propanedithiol, 50, 72
1,3-Propanedithiol di-p-toluenesulfonate and 1,2-eth-
  anedithiol di-p-toluenesulfonate, 54, 33
2-Propanol, 1,3-bis(methylthio)-, 54, 19
Propargyl bromide, coupling
  with an organocopper reagent, 50, 98
Propene, 2-acetoxy-1-phenyl-
  trans-, 54, 50
  trans- and cis-, GLC determination of, 54, 53
Propene, 1,3-bis(methylthio)-,
  54, 24
Propionaldehyde, reaction with
  1,3-bis(methylthio)allyl-
  lithium, 54, 21
Propionaldehyde, 2,2-dimethyl-
  3-phenyl-, 54, 46
PROPIONALDEHYDE, 2-METHYL-3-
  PHENYL-, 51, 17
Propylene, with acetyl chloride,
  aluminum chloride, and quinoline to give trans-3-penten-

2-one, 51, 115
  with acetyl chloride and aluminum chloride to give 4-
  chloropentane-2-one, 51, 116
N-Protection of amino acids,
  53, 27
4H-Pyran, 5,6-dihydro-6-pyrro-
  lidino-2,5,5-trimethyl-,
  53, 49
Pyridine, 1,2,3,4-tetrahydro-1-
  methyl-2-phenyl-, 54, 93
α-Pyrone, irradiation of, 50,
  23
Pyrrole, N-alkylation of, 54,
  59
  reaction with trichloroacetyl
  chloride to give pyrrol-2-
  yl trichloromethyl ketones,
  51, 100
Pyrrole-2-carboxylic acid esters, from pyrrol-2-yl trichloromethyl ketone, 51, 102
PYRROLE-2-CARBOXYLIC ACID,
  ETHYL ESTER, 51, 100
Pyrrolidine, 54, 40
1,2-Pyrrolidinedicarboxylic
  acid, 1-tert-butyl ester,
  L-, 53, 25
Pyrrolidine, 1-(2-methylpro-
  penyl)-, 53, 48
1-Pyrrolidinocyclohexene, preparation of, 54, 40
Pyrrol-2-yl-trichloromethyl ketone, with ethanol to give
  ethyl pyrrole-2-carboxylate,
  51, 100

QUADRICYCLANE, 51, 133
Quadricyclane, preparation of
  substituted derivatives, 51,
  135
  reactions of, 51, 135
Quinoline, with 4-chloropen-
  tane-2-one to give trans-3-
  penten-2-one, 51, 116

Raney nickel alloy, reduction
  of aromatic nitriles to aldehydes, 51, 22
REACTION OF ARYL HALIDES WITH

π-ALLYLNICKEL HALIDES: METHALLYLBENZENE, **52**, 115
Rearrangement of epoxides to allylic alcohols, **53**, 17
Reduction, by controlled-potential electrolysis, **52**, 22
  by lithium aluminum hydride of exo-3,4-dichlorobicyclo[3.2.1]oct-2-ene to 3-chlorobicyclo[3.2.1]oct-2-ene, **51**, 61
  by sodium borohydride of erythro-3-methanesulfonyloxy-2-butyl cyclobutanecarboxylate, **51**, 12
  by sodium borohydride of 2-(1-phenylcyclopentyl)-4,4,6-trimethyl-5,6-dihydro-1,3(4H)-oxazine, **51**, 25
  of acid chlorides with palladium-on-carbon catalyst to give aldehydes, **51**, 10
  of aromatic nuclei, **51**, 105
  of p-cyanobenzenesulfonamide with Raney nickel alloy or p-formylbenzenesulfonamide, **51**, 20
  with aluminum amalgam, **52**, 77
  with chromium(II) salts, **52**, 62
  with hydroxylamine, **52**, 128
  with sodium borohydride, **52**, 122
  with sodium cyanoborohydride, **52**, 124
  with sodium-liquid ammonia, **54**, 11
  with sulfur dioxide, **52**, 83, 135
REDUCTION OF ALKYL HALIDES AND TOSYLATES WITH SODIUM CYANOBOROHYDRIDE IN HEXAMETHYLPHOSPHORAMIDE (HMPA): A. 1-IODODECANE TO n-DECANE; B. 1-DODECYL TOSYLATE TO n-DODECANE, **53**, 107
REDUCTION OF KETONES BY USE OF TOSYLHYDRAZONE DERIVATIVES: ANDROSTAN-17β-OL, **52**, 122
REDUCTIVE AMINATION WITH SODIUM CYANOBOROHYDRIDE: N,N-DIMETHYLCYCLOHEXYLAMINE, **52**, 124
Reference electrode for electrolytic reduction, **52**, 28
Resorcinol dimethyl ether, **50**, 52
Rhodium-on-alumina, catalyzed reduction of aromatic nuclei, **51**, 105
Rosenmund reduction, 3,4,5-trimethoxybenzaldehyde, **51**, 8

Sebacid acid dinitrile, **50**, 20
SELECTIVE α-BROMINATION OF AN ARALKYL KETONE WITH PHENYLTRIMETHYL AMMONIUM TRIBROMIDE: 2-BROMOACETYL-6-METHOXYNAPHTHALENE AND 2,2-DIBROMOACETYL-6-METHOXYNAPHTHALENE, **53**, 111
Shikimic acid, **50**, 27
Silver benzoate, as catalyst in decomposition of diazoketones, **50**, 78
Silver isocyanate, with iodine to give iodine isocyanate, **51**, 112
Sodium, with 1-bromo-3-chlorocyclobutane to give bicyclo[1.1.0]butane, **51**, 55
Sodium amalgam, **50**, 50, 51
Sodium amide, with 2,4-pentanedione and diphenyliodonium chloride to give 1-phenyl-2,4-pentanedione, **51**, 128
Sodium azide, **50**, 107
  with mixed carboxylic-carbonic anhydrides, **51**, 49
Sodium borohydride, reduction of erythro-3-methanesulfonyloxy-2-butyl cyclobutanecarboxylate, **51**, 12
  reduction of 2-(1-phenylcyclopentyl)-4,4,6-trimethyl-5,6-dihydro-1,3(4H)-oxazine to 2-(1-phenylcyclopentyl)-4,4,6-trimethyltetrahydro-1,3-oxazine, **51**, 25
Sodium cyanoborohydride, used

in reductions, 52, 124
Sodium cyanoborohydride-hexamethylphosphoramide, 53, 109
Sodium deuteroxide, 53, 41
Sodium formate, reaction with acetyl chloride, 50, 1
Sommelet reaction, 50, 71
Spiropentane, 52, 32
 from pentaerythrityltetrabromide and sodium, 51, 58
Spiropentanes, 54, 32
Steam distillation, of volatile aldehydes, 51, 33, 36
STEREOSELECTIVE SYNTHESIS OF TRISUBSTITUTED OLEFINS: ETHYL 4-METHYL-E-4,8-NONADIENOATE, 53, 116
Styrene, reaction with carbethoxycarbene, 50, 94
Suberoyl chloride, 54, 88
SUBSTITUTION OF ARYL HALIDES WITH COPPER(I) ACETYLIDES: 2-PHENYL[3,2-b]PYRIDINE, 52, 128
Succinic acid mononitrile, ethyl ester, 50, 20
Succinic anhydride, 52, 1
Sulfide, diphenyl-, 54, 28
Sulfides, aromatic, 50, 76
3-Sulfolene, as a source of 1,3-butadiene *in situ*, 50, 43
Sulfones, bromination, 50, 31
Sulfonium, 3-chloropropyldiphenyl-, fluoroborate, 54, 28
Sulfonium, cyclopropyldiphenyl tetrafluoroborate, 54, 28
Sulfonium salts, acetylenic, furans from, 53, 3
Sulfonium ylides, 54, 32
Sulfur, reaction with organolithium compounds, 50, 105
Sulfuryl chloride, with 1,1-cyclobutanedicarboxylic acid to give 3-chloro-1,1-cyclobutanedicarboxylic acid, 51, 73

Tetracyanomethylene oxide, with n-butyl sulfide to give carbonyl cyanide, 51, 70
2-Tetradecyl-sym-trithiane, by reaction of 1-bromotetradecane with sym-trithiane in presence of n-butyllithium, 51, 39
Tetraethylammonium tetrafluoroborate, 52, 29
1,2,3,4-TETRAHYDRO-β-CARBOLINE, 51, 136
1,2,3,4-Tetrahydro-β-carboline, synthesis of substituted derivatives, 51, 138
1,2,3,4-TETRAHYDRONAPHTHALENE (1,2)IMINE, 51, 53
β-TETRALONE, 6-METHOXY-, 51, 109
β-Tetralones, general synthesis of substituted derivatives, 51, 111
1,1,3,3-Tetramethylbutyl isonitrile, from N-(1,1,3,3-tetramethyl butyl)formamide and thionyl chloride, 51, 31
2,4,4,6-Tetramethyl-5,6-dihydro-1,3(4H)-oxazine, for synthesis of substituted acetaldehydes, 51, 30
2,2,3,3-TETRAMETHYLIODOCYCLOPROPANE, 52, 132
Thermolysis, 1-phenylcyclopentanecarboxylic acid azide to 1-phenyl-cyclopentyl isocyanate, 51, 49
4H-1,4-THIAZINE 1,1-DIOXIDE, 52, 135
2-Thienyllithium, 50, 104
THIIRENE 1,1-DIOXIDE, DIPHENYL-, 50, 65
2,2'-THIODITHIOPHENE, 50, 75
Thiols, general synthetic method, 50, 106
Thiophene, cyanation, 50, 54
2-THIOPHENETHIOL, 50, 75, 104
3-Thiophenethiol, 50, 106
THIOPHENOLS FROM PHENOLS: 2-NAPHTHALENETHIOL, 51, 139
Titanium tetrachloride, 54, 93
o-Tolualdehyde, by reduction of

o-tolunitrile with Raney nickel alloy in formic acid, 51, 23
p-Toluenesulfonyl azide, with 2-(hydroxymethylene) cyclohexanone to give 2-diazocyclohexanone, 51, 86
p-Toluenethiosulfonic acid, potassium salt, 54, 33
p-TOLUIC ACID, tert-BUTYL ESTER, 51, 96
p-Toluoyl chloride, with tert-butanol and n-butyllithium to give tert-butyl p-toluate, 51, 96
Tosylates, reduction of, 53, 109
Tosylhydrazones, formation, 52, 122
 reduction, 52, 122
 with methyllithium to give olefins, 51, 69
Trialkyloxonium salts, as alkylating agents, 51, 144
Triazoles, general route to, 50, 109
1,2,4-TRIAZOLINE-3,5-DIONE, 4-PHENYL-, 51, 121
Tri-n-butylcarbinol, 52, 22
TRI-tert-BUTYLCYCLOPROPENYL FLUOROBORATE, 54, 97
Trichloroacetyl chloride, with pyrrole to give pyrrol-2-yl trichloromethyl ketone, 51, 100
Tricyclo[3.3.1.1^{3,7}]decanone, 53, 8
Tricyclo[4.4.1.0^{1,6}]undeca-3,8-diene, 54, 13
Tricyclo[4.4.1.0^{1,6}]undeca-3,8-diene, 11,11-dichloro-, 54, 12
Triethylaluminum, apparatus and procedures for handling, 52, 90, 96, 100
Triethylamine, in synthesis of diazoketones, 50, 77
Triflates, vinyl-, 54, 82
3,4,5-TRIMETHOXYBENAZLDEHYDE, 51, 8

3,4,5-Trimethoxybenzoyl chloride, reduction to 3,4,5-trimethoxy benazaldehyde, 51, 8
TRIMETHYL-p-BENZOQUINONE, 52, 83
Trimethylchlorosilane, 50, 107
Trimethylcyclohexanones, reduction of axial alcohols, 50, 15
Trimethylene dibromide, 54, 34
2,2-(TRIMETHYLENEDITHIO)CYCLOHEXANONE, 54, 39
Trimethylene dithiotosylate, 54, 40
TRIMETHYLENE DITHIOTOSYLATE AND ETHYLENE DITHIOTOSYLATE, 54, 33
Trimethyleneimine, 53, 13
N,4,4-Trimethyl-2-oxazolinium iodide, preparation of, 54, 44
TRIMETHYLOXONIUM TETRAFLUOROBORATE, 51, 142
Trimethyloxonium tetrafluoroborate, reactions of, 51, 144
2,4,4-TRIMETHYLPENTANAL, 51, 4
TRIMETHYLSILYL AZIDE, 50, 107
Triphenylphosphine-cobalt(II) bromide, catalyst, 53, 30, 32
Triphenylphosphine imine, 50, 83
Triphenyl phosphite, with methyl iodide and cyclohexanol, 51, 45
 with neopentyl alcohol and methyl iodide, 51, 44
Tris(triphenylphosphine)rhodium chloride, 53, 64
sym-Trithiane, reaction with 1-bromotetradecane in presence of n-butyllithium, 51, 39
Tryptamine, with glyoxylic acid to give 1,2,3,4-tetrahydro-β-carboline, 51, 136

cis-α,β-UNSATURATED ACIDS: ISOCROTONIC ACID, 53, 123

α,β-Unsaturated carbonyl compounds, preparative method, **50**, 70

Vacuum manifold system, **51**, 56
Vanadium oxyacetylacetonate, **50**, 56
N-Vinylpyrrolidone, cyanation, **50**, 54

Wittig reaction, **53**, 104

Wurtz reaction, bicyclo[1.1.0] butane from 1-bromo-3-chlorocyclo butane, **51**, 55

Zinc, cyclopropane from 1,3-dichloropropane, **51**, 58
Zinc, activated, **53**, 88
Zinc chloride, anhydrous, ethereal solution, preparation of, **54**, 54

Unchecked Procedures

Received during the period July 10, 1973 - July 1, 1974

In accordance with a policy adopted by the Board of Editors beginning with Volume 50, which, as noted in the Editor's Preface, is intended to make procedures available more rapidly, procedures received by the Secretary during the year, whether subsequently accepted, modified, or rejected for publication by Organic Syntheses, will be made available for purchase at the price of $2 per procedure, prepaid, upon request to the Secretary:

> Dr. Wayland E. Noland, Secretary
> Organic Syntheses
> Department of Chemistry
> University of Minnesota
> Minneapolis, Minnesota 55455

Payment must accompany the order, and should be made payable to Organic Syntheses, Inc. (not to the Secretary). Purchase orders not accompanied by payment will not be accepted. Procedures may be ordered by number and/or title from the list which follows.

It should be emphasized that the procedures which are being made available are unedited and have been reproduced just as they are first received from the submitters. There is no assurance that the procedures listed here will utimately check in the form available, and many of them have been or will be rejected for publication in Organic Syntheses either before or after the checking process. For this reason, Organic Syntheses can provide no assurance whatsoever that the procedures will work as described, and offers no comment as to what safety hazards may be involved. Consequently, more than usual caution should be employed in following the directions in the procedures.

Organic Syntheses welcomes, on a strictly voluntary basis, comments from persons who attempt to carry out the procedures. For this purpose, a Checker's Report form will be mailed out with each unchecked procedure ordered. Procedures which have been checked by or under the supervision of a member of the Board of Editors will continue to be published in the volumes of Organic Syntheses, as in the past. It is anticipated that many of the procedures in the list will be published (often in revised form) in Organic Syntheses in future volumes.

-Wayland E. Noland

1890 <u>3-Alkylated and 3-Acylated Indoles Through the Preparation and Use of 2-Methylthio-1,3-dithiane Derivatives. 3-Benzoylindole and 3-Benzylindole</u>

P. Stütz and P. A. Stadler, Pharmaceutical Division, Chemical Research, Sandoz Ltd., Basel, Switzerland

1891   A Synthesis of 5-Acetyl-1,2,3,4,5-pentamethylcyclo-
        pentadiene

R. B. King, W. M. Douglas, and A. Efraty, Department of Chemistry, University of Georgia, Athens, Ga. 30601

1892   1-Benzylisoquinoline Formation via N-Benzoyl-1,2-dihydroisoquinaldonitrile

Barrie C. Uff, John R. Kershaw, and John L. Neumeyer, Department of Chemistry, Loughborough University of Technology, Loughborough, Leicestershire, LE11 3TU, England, U.K.

$$\text{isoquinoline} + \emptyset COCl + KCN \xrightarrow[H_2O]{CH_2Cl_2} \text{N-benzoyl-1,2-dihydroisoquinaldonitrile (NCO}\emptyset\text{, CN)}$$
64-69%

$$\xrightarrow[\substack{DMF \\ 0°}]{\substack{NaH \\ \emptyset CH_2Cl}} [\text{intermediate with NCO}\emptyset, NC, CH_2\emptyset] \xrightarrow[\substack{H_2O \\ reflux}]{\substack{NaOH \\ EtOH}} \text{1-benzylisoquinoline (CH}_2\emptyset\text{)}$$
75-84%

1893   Isonitriles by the Phase Transfer Hofmann Carbylamine Reaction: tert-Butyl Isocyanide

George W. Gokel, Ronald P. Widera, and William P. Weber, Department of Chemistry, University of Southern California, University Park, Los Angeles, Calif. 90007

$$(CH_3)_3CNH_2 + CHCl_3 \xrightarrow[\substack{H_2O, CH_2Cl_2 \\ reflux}]{\substack{NaOH \\ + \\ \emptyset CH_2NEt_3\ Cl^{\ominus}}} (CH_3)_3CN=C$$
66-73%

1894 5-Hydroxy-2-methylpyridine

A. U. De and B. P. Saha, Department of Pharmacy, Jadavpur University, Calcutta 32, India

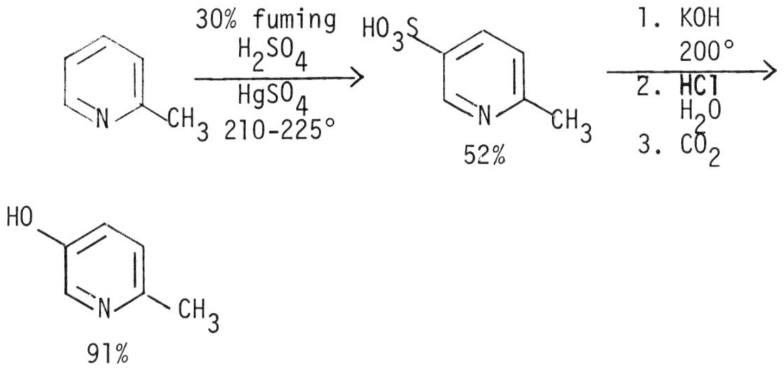

$\underset{\text{CH}_3}{\text{pyridine}} \xrightarrow[\text{HgSO}_4 \; 210\text{-}225°]{\text{30\% fuming H}_2\text{SO}_4} \underset{52\%}{\text{HO}_3\text{S-pyridine-CH}_3} \xrightarrow[\substack{2.\ \text{HCl} \\ \text{H}_2\text{O} \\ 3.\ \text{CO}_2}]{1.\ \text{KOH}\ 200°}$

HO–pyridine–CH$_3$
91%

1895 Tropolone

Richard A. Minns, Department of Chemistry, Harvard University, Cambridge, Mass. 02138

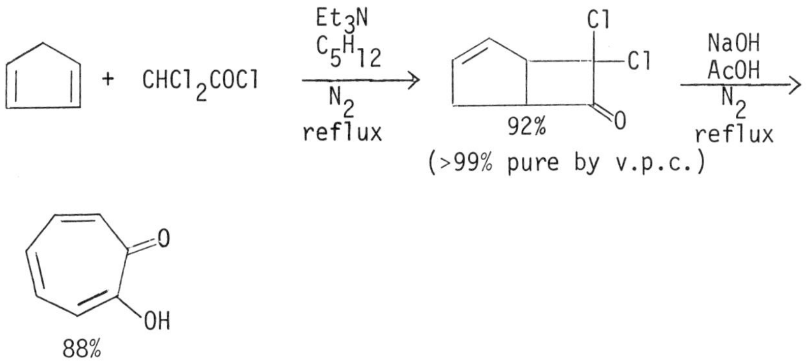

cyclopentadiene + CHCl$_2$COCl $\xrightarrow[\substack{\text{N}_2 \\ \text{reflux}}]{\text{Et}_3\text{N} \\ \text{C}_5\text{H}_{12}}$ [bicyclic dichloroketone] 92% (>99% pure by v.p.c.) $\xrightarrow[\substack{\text{N}_2 \\ \text{reflux}}]{\text{NaOH} \\ \text{AcOH}}$

tropolone
88%

1896   Methyl 2-Alkynoates from 3-Alkyl or 3-Aryl-2-pyrazolin-5-ones: Methyl 2-Hexynoate

Edward C. Taylor, Roger L. Robey, David K. Johnson, and Alexander McKillop, Department of Chemistry, Princeton University, Princeton, N. J. 08540

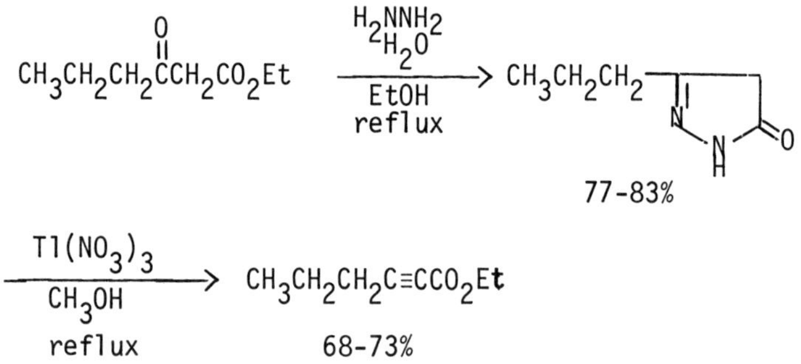

$$CH_3CH_2CH_2\underset{O}{\overset{\parallel}{C}}CH_2CO_2Et \xrightarrow[\text{EtOH reflux}]{H_2NNH_2 \cdot 2H_2O} CH_3CH_2CH_2\text{-pyrazolinone}$$

77-83%

$$\xrightarrow[\text{CH}_3\text{OH reflux}]{Tl(NO_3)_3} CH_3CH_2CH_2C\equiv CCO_2Et$$

68-73%

1897 <u>2-Bromofluorene</u>

David Bromley, Alexander McKillop, and Edward C. Taylor,
Department of Chemistry, Princeton University, Princeton, N. J.
08540

$$\text{fluorene} \xrightarrow[\substack{CCl_4, N_2 \\ \text{reflux}}]{\substack{Tl(OCOCH_3)_3 \\ Br_2}} \text{2-bromofluorene}$$

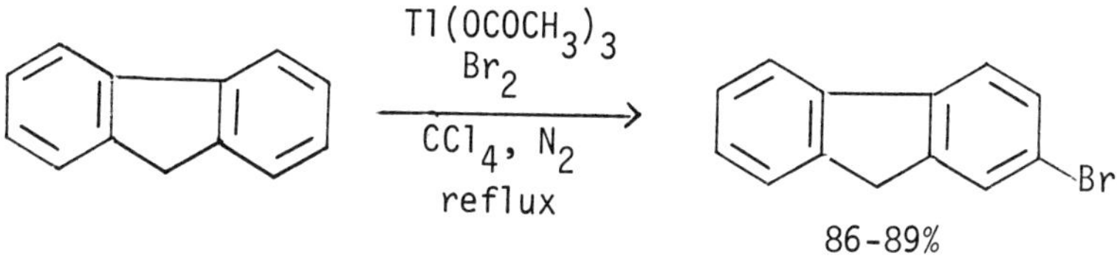

86-89%

1898 <u>2-Iodo-p-xylene</u>

Edward C. Taylor, Frank Kienzle, and Alexander McKillop,
Department of Chemistry, Princeton University, Princeton, N. J.
08540

$$p\text{-xylene} \xrightarrow[CF_3CO_2H]{Tl(OCOCF_3)} \left[\text{ArTl(OCOCF_3)}_2 \right] \xrightarrow[\substack{H_2O \\ Et_2O}]{KI} \text{2-iodo-p-xylene}$$

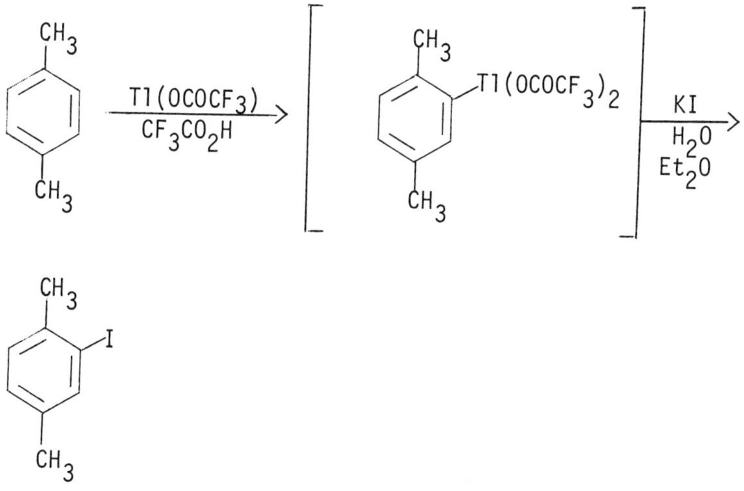

80-84%

1899 3-Bromoacetophenone

Richard F. Eizember and Ann S. Ammons, Process Research and Development, The Lilly Research Laboratories, Eli Lilly and Company, Indianapolis, Ind. 46206

$$\text{PhCOCH}_3 \xrightarrow[\text{ClCH}_2\text{CH}_2\text{Cl}]{\text{Br}_2, \text{AlCl}_3, 30\text{-}40°} \xrightarrow[0°]{\text{H}_2\text{O}} \text{3-BrC}_6\text{H}_4\text{COCH}_3 \quad 70\text{-}74\%$$

1900 **Ethyl 4-Amino-3-methylbenzoate**

P. G. Gassman and G. Gruetzmacher, Department of Chemistry, Ohio State University, Columbus, O. 43210

1901   Ethyl 2-Methylindole-5-carboxylate

P. G. Gassman and T. J. van Bergen, Department of Chemistry,
Ohio State University, Columbus, O.   43210

$$CH_3COCH_2Cl \xrightarrow[\substack{CH_3OH \\ 0-5°}]{\substack{CH_3SH \\ CH_3ONa}} CH_3COCH_2SCH_3 \quad 72\%$$

$$EtOCO\text{-}C_6H_4\text{-}NH_2 \xrightarrow[\substack{CH_2Cl_2 \\ N_2 \\ -70°}]{(CH_3)_3COCl} \left[EtOCO\text{-}C_6H_4\text{-}NHCl \right]$$

$$\xrightarrow[\substack{CH_2Cl_2 \\ -70°}]{CH_3COCH_2SCH_3} \left[EtOCO\text{-}C_6H_4\text{-}NH\text{-}S^+(CH_3)(CH_2COCH_3) \; Cl^- \right] \xrightarrow[CH_2Cl_2]{Et_3N}$$

EtOCO-[5-indole]-3-SCH$_3$, 2-CH$_3$ 51-61%

$$\xrightarrow[EtOH]{Raney\;Ni}$$

EtOCO-[5-indole]-2-CH$_3$ 93%

142

7-Methyloxindole

P. G. Gassman and T. J. van Bergen, Department of Chemistry, Ohio State University, Columbus, O. 43210

$ClCH_2CO_2Et \xrightarrow[\text{EtOH}]{\text{CH}_3\text{SH, Na}}_{0-5°} CH_3SCH_2CO_2Et$ (77%)

[Reaction scheme: o-toluidine + (CH₃)₃COCl in CH₂Cl₂, N₂, −70° → N-chloro intermediate; then CH₃SCH₂CO₂Et, CH₂Cl₂, −70° → sulfonium salt intermediate; then 1. Et₃N, CH₂Cl₂; 2. 2 N HCl, Et₂O → 3-methylthio-7-methyloxindole (63–71%)]

Raney Ni / EtOH, reflux → 7-methyloxindole (70%)

1903   4,4'-Dimethylbiphenyl

L. F. Elsom, Alexander McKillop, and Edward C. Taylor, Department of Chemistry, Princeton University, Princeton, N. J. 08540

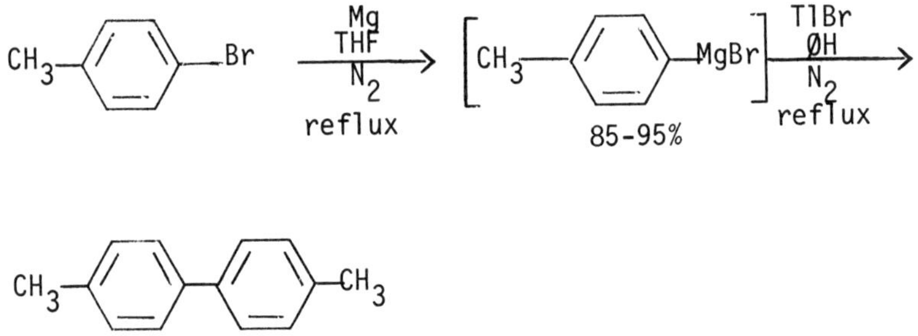

$$CH_3-C_6H_4-Br \xrightarrow[\text{reflux}]{\text{Mg, THF, N}_2} [CH_3-C_6H_4-MgBr] \xrightarrow[\text{reflux}]{\text{TlBr, OH, N}_2} $$

$$CH_3-C_6H_4-C_6H_4-CH_3$$

83-92%
(from p-bromotoluene)

1904   Ferrocenecarboxylic Acid

Perry C. Reeves, Department of Chemistry, Southern Methodist University, Dallas, Tex. 75222

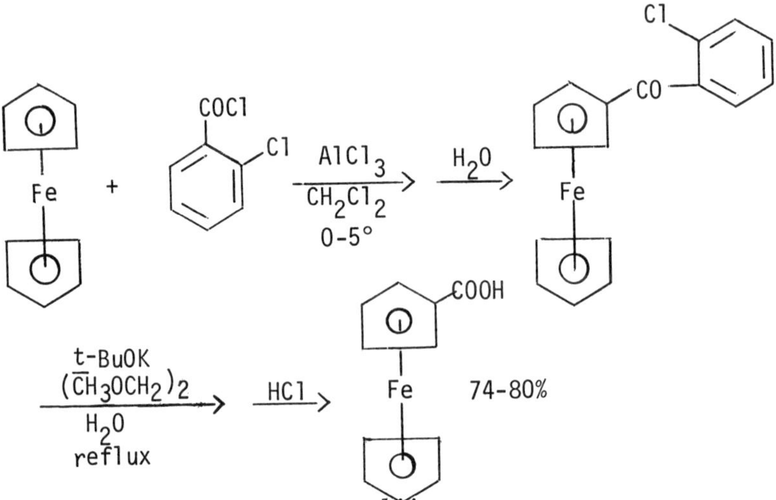

Fe + o-Cl-C$_6$H$_4$-COCl $\xrightarrow[\text{CH}_2\text{Cl}_2, 0-5°]{\text{AlCl}_3}$ $\xrightarrow{\text{H}_2\text{O}}$ Fe-CO-C$_6$H$_4$-Cl

$\xrightarrow[\text{H}_2\text{O, reflux}]{\text{t-BuOK, (CH}_3\text{OCH}_2)_2}$ $\xrightarrow{\text{HCl}}$ Fe-COOH   74-80%

1905 Polymeric Carbodiimide Reagent

Ned M. Weinshenker, Chah M. Shen, and Jack Y. Wong, Dynapol, 1454 Page Mill Road, Palo Alto, Calif. 94304

$(-\underset{|}{C}HCH_2-)_n$–C₆H₅ $\xrightarrow[\text{CHCl}_3, 0°]{ClCH_2OCH_3, SnCl_4}$ $(-\underset{|}{C}HCH_2-)_n$–C₆H₄–CH₂Cl $\xrightarrow[\text{DMF, 100°}]{\text{phthalimide-NK}}$

$\xrightarrow[\text{reflux}]{H_2NNH_2, EtOH}$ $(-\underset{|}{C}HCH_2-)_n$–C₆H₄–CH₂NH₂ $\xrightarrow[\text{THF}]{i\text{-PrNCO}}$ $(-\underset{|}{C}HCH_2-)_n$–C₆H₄–CH₂NHC(O)NH-$i$-Pr

$\xrightarrow[\text{CH}_2\text{Cl}_2 \text{ reflux}]{TsCl, Et_3N}$ $(-\underset{|}{C}HCH_2-)_n$–C₆H₄–CH₂N=C=N-$i$-Pr

Secondary and Tertiary Alkyl Ketones from Carboxylic Acid Chlorides and Lithium Phenylthio(alkyl)cuprate(I) Reagents: Pivalophenone

Gary H. Posner and Charles E. Whitten, Department of Chemistry, The Johns Hopkins University, Baltimore, Md. 21218

$$\phi SLi + (CH_3)_3CLi \xrightarrow[\text{THF}]{\text{CuI} \; N_2} \phi S[(CH_3)_3C]CuLi$$

$$\text{PhCOCl} \xrightarrow[-78°]{\text{THF}} \text{PhCOC(CH}_3)_3$$

85-87%

1907 **2,2,7,7,12,12,17,17-Octamethyl-21,22,23,24-tetra-oxaperhydroquaterene**

Maurice Chastrette, Francine Chastrette, and Jean Sabadie, Laboratoire de Chimie Organique Physique, Université Claude-Bernard Lyon I, 43, Boulevard du 11 Novembre 1918, 69 - Villeurbanne, France

furan + CH_3COCH_3 $\xrightarrow[\text{EtOH} \atop 60\text{-}65°]{\text{HCl} \atop \text{LiClO}_4}$ octamethyltetraoxaquaterene (85-95%)

$\xrightarrow[\text{EtOH} \atop 105°]{170 \text{ bars } H_2 \atop 10\% \text{ Pd-C}}$ perhydro product (46%)

1908 Esterification of Carboxylic Acids with Trialkyloxonium Salts: Ethyl and Methyl 4-Acetoxybenzoates

Douglas J. Raber, Patrick Gariano, Jr., Albert O. Brod, Anne L. Gariano, and Wayne C. Guida, Department of Chemistry, University of South Florida, Tampa, Fla. 33620

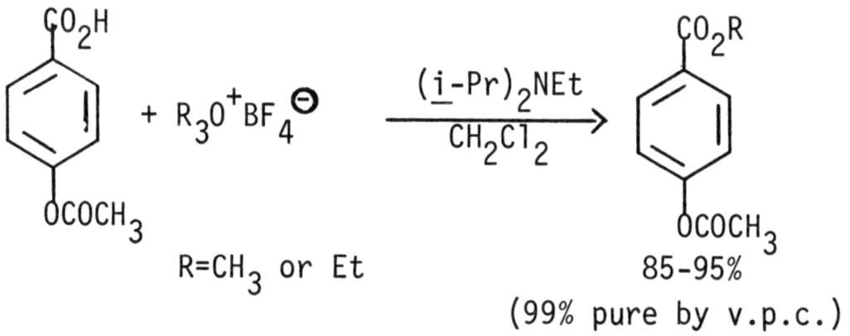

$$\text{4-CH}_3\text{COO-C}_6\text{H}_4\text{-CO}_2\text{H} + R_3\text{O}^+\text{BF}_4^- \xrightarrow[\text{CH}_2\text{Cl}_2]{(\underline{i}\text{-Pr})_2\text{NEt}} \text{4-CH}_3\text{COO-C}_6\text{H}_4\text{-CO}_2\text{R}$$

R = CH$_3$ or Et   85-95%
                         (99% pure by v.p.c.)

1909 trans-Ethyl 3-Nitroacrylate

John E. McMurry and John H. Musser, Natural Sciences I, University of California, Santa Cruz, Santa Cruz, Calif. 95060

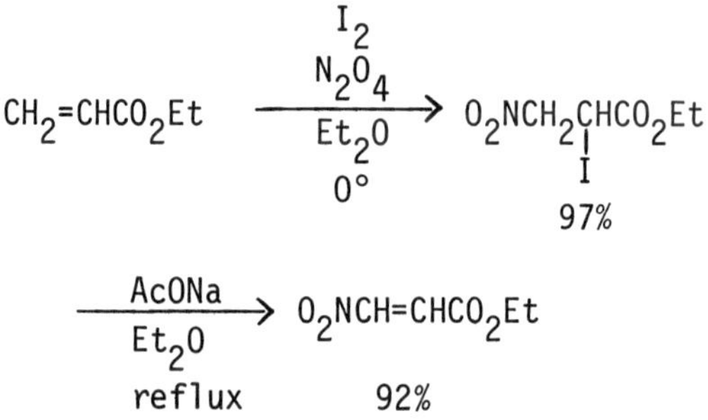

$$\text{CH}_2\text{=CHCO}_2\text{Et} \xrightarrow[\substack{\text{Et}_2\text{O} \\ 0°}]{\substack{\text{I}_2 \\ \text{N}_2\text{O}_4}} \text{O}_2\text{NCH}_2\text{CHICO}_2\text{Et}$$

97%

$$\xrightarrow[\substack{\text{Et}_2\text{O} \\ \text{reflux}}]{\text{AcONa}} \text{O}_2\text{NCH=CHCO}_2\text{Et}$$

92%

1910 **3,5,5-Trimethyl-2-(2-oxopropyl)-2-cyclohexen-1-one**

Z. Valenta and H. J. Liu, Department of Chemistry, University of Alberta, Edmonton, Alberta, Canada T6G 2E1

2,3,4,5,6-Pentafluorophenylacetonitrile and its Conversion to 4,5,6,7-Tetrafluoroindole via 2-Pentafluorophenylethylamine

Robert Filler and Sarah M. Woods, Department of Chemistry, Illinois Institute of Technology, Chicago, Ill. 60616

1912 **Methyl 2-Oxocyclohexanecarboxylate**

P. Deslongchamps and L. Ruest, Laboratoire de synthese organique, Département de chimie, Université de Sherbrooke, Sherbrooke, Quebec, Canada JIK 2R1

$$\text{cyclohexanone} + CH_3OCO_2CH_3 \xrightarrow[\text{THF reflux}]{\text{NaH, KH}} \xrightarrow[H_2O]{\text{AcOH}} \text{methyl 2-oxocyclohexanecarboxylate (93\%)}$$

1913 **Biaryls from Simple Arenes, via Telluriumorganic Reagents: 4,4'-Dimethoxybiphenyl**

J. Bergman, R. Carlsson, and B. Sjöberg, Department of Organic Chemistry, Royal Institute of Technology, S-100 44 Stockholm 70, Sweden

$$CH_3O\text{-}C_6H_5 \xrightarrow[160°]{TeCl_4} CH_3O\text{-}C_6H_4\text{-}TeCl_2\text{-}C_6H_4\text{-}OCH_3 \quad (84\text{-}90\%)$$

$$\xrightarrow[\substack{N_2 \\ 200°}]{\text{Raney Ni}} CH_3O\text{-}C_6H_4\text{-}C_6H_4\text{-}OCH_3 \quad (78\text{-}90\%)$$

1914   7-Hydroxyindan-1-one

H.J.E. Loewenthal and S. Schatzmiller, Chemistry Department,
Israel Institute of Technology, Haifa, Israel

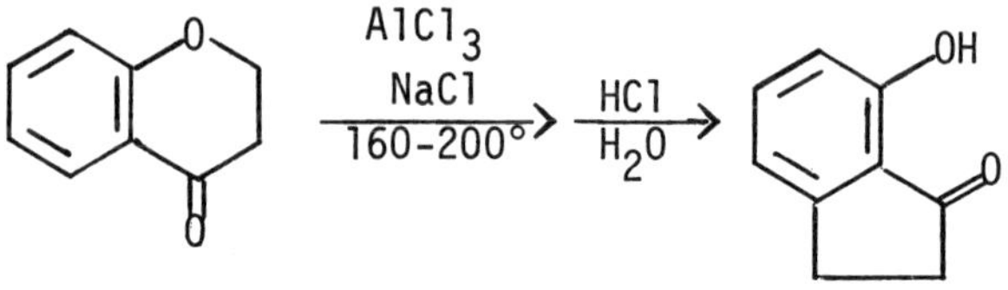

chromanone $\xrightarrow[160-200°]{\text{AlCl}_3, \text{NaCl}} \xrightarrow[\text{H}_2\text{O}]{\text{HCl}}$ 7-hydroxyindan-1-one

68%

1915   Biguanide Di-p-toluenesulfonate

C. P. Joshua and V. P. Rajan, Department of Chemistry, University
of Kerala, Trivandrum-695001, India

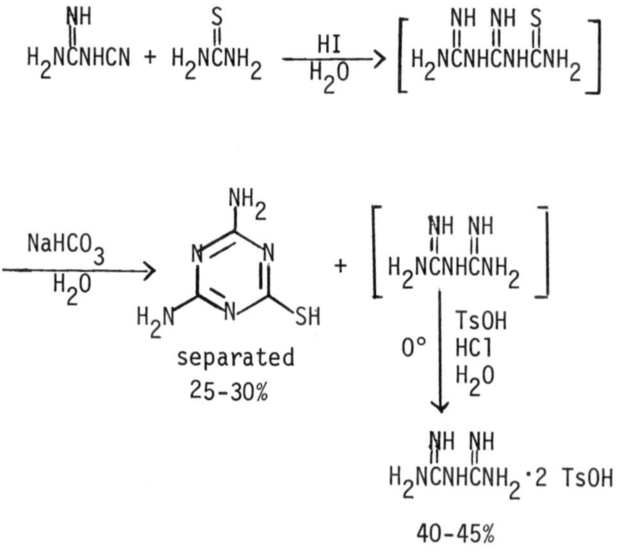

$$H_2NCNHCN + H_2NCNH_2 \xrightarrow[H_2O]{HI} [H_2NCNHCNHCNH_2]$$

$\xrightarrow[H_2O]{NaHCO_3}$ 2-amino-4-amino-6-mercapto-1,3,5-triazine + $[H_2NCNHCNH_2]$

separated 25-30%

$\xrightarrow[H_2O]{0°, TsOH, HCl}$

$H_2NCNHCNH_2 \cdot 2\ TsOH$

40-45%

1916 Sulfonyl Cyanides: Benzenesulfonyl Cyanide

M. S. A. Vrijland, Technische Hogeschool Twente, Postbus 217, Enschede, The Netherlands

$$\emptyset SO_2Cl \xrightarrow[H_2O]{\substack{Na_2SO_3 \\ NaHCO_3}} [\emptyset SO_2Na] \xrightarrow[10°]{ClCN} \emptyset SO_2CN$$
$$82-86\%$$

1917 Benzyl N-(1-Chloro-2,2,2-trifluoroethyl)carbamate via Benzyl N-(2,2,2-Trifluoro-1-hydroxyethyl)carbamate

Per Larsen, The Danish Institute of Protein Chemistry, 4, Venlighedsvej, DK-2970 Hørsholm, Denmark

$$CF_3CO_2Et \xrightarrow[-75° \text{ to } -70°]{LiAlH_4 \\ THF} \left[(CF_3\overset{\overset{OEt}{|}}{C}HO)_4AlLi \right] \xrightarrow[reflux]{\substack{H_2NCO_2CH_2\emptyset \\ BF_3 \cdot Et_2O}}$$

$$\xrightarrow{1 \text{ N } HCl} CF_3\overset{\overset{OH}{|}}{C}HNHCO_2CH_2\emptyset \xrightarrow[80°]{\substack{PCl_5 \\ POCl_3}} CF_3\overset{\overset{Cl}{|}}{C}HNHCO_2CH_2\emptyset$$
$$75-78\% \qquad\qquad 90-94\%$$

1918 Cinnamaldehyde from Cinnamyl Alcohol

Udo A. Spitzer and Donald G. Lee, Department of Chemistry, University of Saskatchewan, Regina Campus, Regina, Canada S4S A02

$$\emptyset CH=CHCH_2OH \xrightarrow[\substack{H_2O \\ reflux}]{Na_2Cr_2O_7} \emptyset CH=CHCHO$$
$$70-80\%$$

1919   O-Mesitylenesulfonylhydroxylamine

Y. Tamura, J. Minamikawa, and M. Ikeda, Faculty of Pharmaceutical Sciences, University of Osaka, Toyonaka, Osaka-fu, Japan

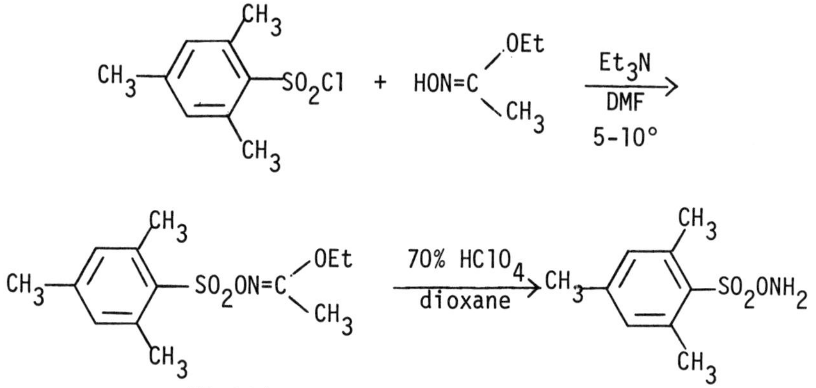

$$\text{Mesityl-SO}_2\text{Cl} + \text{HON=C(OEt)(CH}_3\text{)} \xrightarrow[5-10°]{\text{Et}_3\text{N}, \text{DMF}} \text{Mesityl-SO}_2\text{ON=C(OEt)(CH}_3\text{)} \xrightarrow[\text{dioxane}]{70\% \text{ HClO}_4} \text{Mesityl-SO}_2\text{ONH}_2$$

72-86%

1920   9-Thiabicyclo[4.2.1]nona-2,4-diene

Peter H. McCabe and C. Mary Livingston, Department of Chemistry, The University, Glasgow G12 8QQ, Scotland

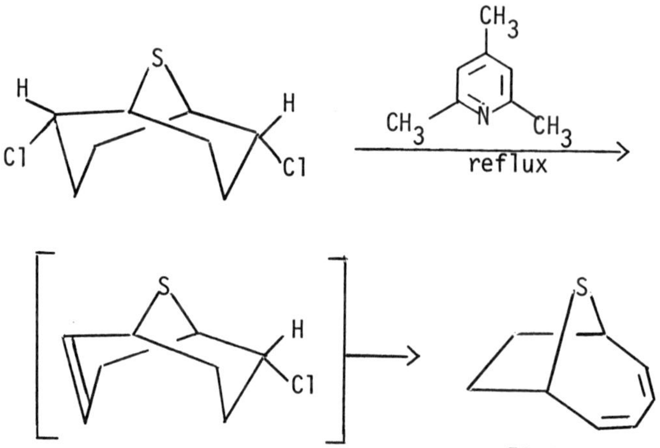

50-65%

1921 Dimethyl Allene-1,3-dicarboxylate

T. A. Bryson and T. Dolak, Department of Chemistry, University of South Carolina, Columbia, S. C. 29208

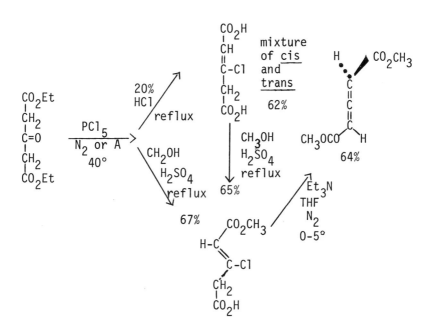

1922   Trifluoroacetylation of Amines and Amino Acids Under Neutral and Mild Conditions: A. N-Trifluoroacetanilides; B. N-Trifluoroacetyl-L-tyrosine

C. A. Panetta, Department of Chemistry, The University of Mississippi, University, Miss. 38677

$$\varnothing NH_2 + CF_3CCCl_3 \xrightarrow[25-35°]{DMSO} \varnothing NHCCF_3 + CHCl_3$$
$$69\%$$

$$HO-\langle\rangle-CH_2CHCO_2H + CF_3CCCl_3 \xrightarrow[25-35°]{DMSO}$$
$$\quad\quad\quad\quad NH_2$$

$$HO-\langle\rangle-CH_2CHCO_2H + CHCl_3$$
$$\quad\quad\quad\quad NHCCF_3$$
72-80%

1923   cis-cis-1,4-Dicyano-1,3-butadiene

Jiro Tsuji and Hiroshi Takayanagi, Department of Chemical Engineering, Faculty of Engineering, Tokyo Institute of Technology, O-okayama, Meguro-ku, Tokyo 152, Japan

o-C₆H₄(NH₂)₂ + O₂ $\xrightarrow[\text{pyridine}]{\text{CuCl}}$ 1,2-C₆H₄(CN)₂

73-77%